D1237294

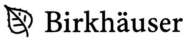

Titu Andreescu • Bogdan Enescu

Mathematical Olympiad Treasures

Second Edition

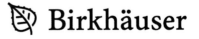 Birkhäuser

Titu Andreescu
School of Natural Sciences and
    Mathematics
University of Texas at Dallas
Richardson, TX 75080
USA
titu.andreescu@utdallas.edu

Bogdan Enescu
Department of Mathematics
"BP Hasdeu" National College
Buzau 120218
Romania
bogdanenescu@buzau.ro

ISBN 978-0-8176-8252-1   e-ISBN 978-0-8176-8253-8
DOI 10.1007/978-0-8176-8253-8
Springer New York Dordrecht Heidelberg London

Library of Congress Control Number: 2011938426

Mathematics Subject Classification (2010): 00A05, 00A07, 05-XX, 11-XX, 51-XX, 97U40

© Springer Science+Business Media, LLC 2004, 2011
All rights reserved. This work may not be translated or copied in whole or in part without the written
permission of the publisher (Springer Science+Business Media, LLC, 233 Spring Street, New York,
NY 10013, USA), except for brief excerpts in connection with reviews or scholarly analysis. Use in
connection with any form of information storage and retrieval, electronic adaptation, computer software,
or by similar or dissimilar methodology now known or hereafter developed is forbidden.
The use in this publication of trade names, trademarks, service marks, and similar terms, even if they are
not identified as such, is not to be taken as an expression of opinion as to whether or not they are subject
to proprietary rights.

Printed on acid-free paper

Springer is part of Springer Science+Business Media (www.birkhauser-science.com)

Preface

Mathematical Olympiads have a tradition longer than one hundred years. The first mathematical competitions were organized in Eastern Europe (Hungary and Romania) by the end of the 19th century. In 1959 the first International Mathematical Olympiad was held in Romania. Seven countries, with a total of 52 students, attended that contest. In 2010, the IMO was held in Kazakhstan. The number of participating countries was 97, and the number of students 517.

Obviously, the number of young students interested in mathematics and mathematical competitions is nowadays greater than ever. It is sufficient to visit some mathematical forums on the net to see that there are tens of thousands registered users and millions of posts.

When we were thinking about writing this book, we asked ourselves to whom it will be addressed. Should it be the beginner student, who is making the first steps in discovering the beauty of mathematical problems, or, maybe, the more advanced reader, already trained in competitions. Or, why not, the teacher who wants to use a good set of problems in helping his/her students prepare for mathematical contests.

We have decided to take the hard way and have in mind all these potential readers. Thus, we have selected Olympiad problems of various levels of difficulty. Some are rather easy, but definitely not exercises; some are quite difficult, being a challenge even for Olympiad experts.

Most of the problems come from various mathematical competitions (the International Mathematical Olympiad, The Tournament of the Towns, national Olympiads, regional Olympiads). Some problems were created by the authors and some are folklore.

The problems are grouped in three chapters: Algebra, Geometry and Trigonometry, and Number Theory and Combinatorics. This is the way problems are classified at the International Mathematical Olympiad.

In each chapter, the problems are clustered by topic into self-contained sections. Each section begins with elementary facts, followed by a number of carefully selected problems and an extensive discussion of their solutions. At the end of each section the reader will find a number of proposed problems, whose complete solutions are presented in the second part of the book.

We encourage the beginning reader to carefully examine the problems solved at the beginning of each section and try to solve the proposed problems before examining the solutions provided at the end of the book. As for the advanced reader, our advice is to try finding alternative solutions and generalizations of the proposed problems.

In the second edition of the book, we added two new sections in Chaps. 1 and 3, and more than 60 new problems with complete solutions.

University of Texas at Dallas   Titu Andreescu
"B.P. Hasdeu" National College   Bogdan Enescu

Contents

Part I
Problems

Chapter 1
Algebra

1.1 An Algebraic Identity

A very useful algebraic identity is derived by considering the following problem.

Problem 1.1 Factor $a^3 + b^3 + c^3 - 3abc$.

Solution Let P denote the polynomial with roots a, b, c:

$$P(X) = X^3 - (a+b+c)X^2 + (ab+bc+ca)X - abc.$$

Because a, b, c satisfy the equation $P(x) = 0$, we obtain

$$a^3 - (a+b+c)a^2 + (ab+bc+ca)a - abc = 0,$$
$$b^3 - (a+b+c)b^2 + (ab+bc+ca)b - abc = 0,$$
$$c^3 - (a+b+c)c^2 + (ab+bc+ca)c - abc = 0.$$

Adding up these three equalities yields

$$a^3 + b^3 + c^3 - (a+b+c)(a^2+b^2+c^2) + (ab+bc+ca)(a+b+c) - 3abc = 0.$$

Hence

$$a^3 + b^3 + c^3 - 3abc = (a+b+c)(a^2+b^2+c^2 - ab - bc - ca). \tag{1.1}$$

Note that the above identity leads to the following result: if $a+b+c = 0$, then $a^3 + b^3 + c^3 = 3abc$.

Another way to obtain the identity (1.1) is to consider the determinant:

$$D = \begin{vmatrix} a & b & c \\ c & a & b \\ b & c & a \end{vmatrix}.$$

T. Andreescu, B. Enescu, *Mathematical Olympiad Treasures*,
DOI 10.1007/978-0-8176-8253-8_1, © Springer Science+Business Media, LLC 2011

Expanding D, we obtain

$$D = a^3 + b^3 + c^3 - 3abc.$$

On the other hand, adding all columns to the first one gives

$$D = \begin{vmatrix} a+b+c & b & c \\ a+b+c & a & b \\ a+b+c & c & a \end{vmatrix} = (a+b+c) \begin{vmatrix} 1 & b & c \\ 1 & a & b \\ 1 & c & a \end{vmatrix}$$

$$= (a+b+c)(a^2 + b^2 + c^2 - ab - bc - ca).$$

Notice that the expression

$$a^2 + b^2 + c^2 - ab - bc - ca$$

can also be written as

$$\frac{1}{2}\left[(a-b)^2 + (b-c)^2 + (c-a)^2\right].$$

We obtain another version of the identity (1.1):

$$a^3 + b^3 + c^3 - 3abc = \frac{1}{2}(a+b+c)\left[(a-b)^2 + (b-c)^2 + (c-a)^2\right]. \quad (1.2)$$

This form leads to a short proof of the $AM - GM$ inequality for three variables. Indeed, from (1.2) it is clear that if a, b, c are positive, then $a^3 + b^3 + c^3 \geq 3abc$. Now, if x, y, z are positive numbers, taking $a = \sqrt[3]{x}, b = \sqrt[3]{y}$ and $c = \sqrt[3]{z}$ yields

$$\frac{x+y+z}{3} \geq \sqrt[3]{xyz},$$

with equality if and only if $x = y = z$.

Finally, let us regard

$$a^2 + b^2 + c^2 - ab - bc - ca$$

as a quadratic in a, with parameters b, c. This quadratic has discriminant

$$\Delta = (b+c)^2 - 4(b^2 + c^2 - bc) = -3(b-c)^2.$$

Hence its roots are

$$a_1 = \frac{b + c - i(b-c)\sqrt{3}}{2} = b\frac{1 - i\sqrt{3}}{2} + c\frac{1 + i\sqrt{3}}{2}$$

and

$$a_2 = \frac{b + c + i(b-c)\sqrt{3}}{2} = b\frac{1 + i\sqrt{3}}{2} + c\frac{1 - i\sqrt{3}}{2}.$$

Setting $\omega = \frac{-1+i\sqrt{3}}{2}$, one of the primitive cubic roots of the unity, we have $\omega^2 = \frac{-1-i\sqrt{3}}{2}$, hence $a_1 = -b\omega - c\omega^2$ and $a_2 = -b\omega^2 - c\omega$.

This gives the factorization

$$a^2 + b^2 + c^2 - ab - bc - ca = \left(a + b\omega + c\omega^2\right)\left(a + b\omega^2 + c\omega\right),$$

which leads to the identity

$$a^3 + b^3 + c^3 - 3abc = (a + b + c)\left(a + b\omega + c\omega^2\right)\left(a + b\omega^2 + c\omega\right). \qquad (1.3)$$

Here are some problems that show how useful the above identities can be.

Problem 1.2 Factor $(x - y)^3 + (y - z)^3 + (z - x)^3$.

Solution Observe that if $a + b + c = 0$, then it follows from (1.1) that $a^3 + b^3 + c^3 = 3abc$. Because

$$(x - y) + (y - z) + (z - x) = 0,$$

we obtain the factorization

$$(x - y)^3 + (y - z)^3 + (z - x)^3 = 3(x - y)(y - z)(z - x).$$

Problem 1.3 Prove that $\sqrt[3]{2 + \sqrt{5}} + \sqrt[3]{2 - \sqrt{5}}$ is a rational number.

Solution Let $x = \sqrt[3]{2 + \sqrt{5}} + \sqrt[3]{2 - \sqrt{5}}$. We then have

$$x - \sqrt[3]{2 + \sqrt{5}} - \sqrt[3]{2 - \sqrt{5}} = 0.$$

As we have seen, $a + b + c = 0$ implies $a^3 + b^3 + c^3 = 3abc$, so we obtain

$$x^3 - \left(2 + \sqrt{5}\right) - \left(2 - \sqrt{5}\right) = 3x\sqrt[3]{\left(2 + \sqrt{5}\right)\left(2 - \sqrt{5}\right)},$$

or

$$x^3 + 3x - 4 = 0.$$

Clearly, one of the roots of this equation is $x = 1$ and the other two roots satisfy the equation $x^2 + x + 4 = 0$, which has no real solutions. Since $\sqrt[3]{2 + \sqrt{5}} + \sqrt[3]{2 - \sqrt{5}}$ is a real root, it follows that

$$\sqrt[3]{2 + \sqrt{5}} + \sqrt[3]{2 - \sqrt{5}} = 1,$$

which is a rational number.

Next come some proposed problems.

Problem 1.4 Factor $(a + 2b - 3c)^3 + (b + 2c - 3a)^3 + (c + 2a - 3b)^3$.

Problem 1.5 Let x, y, z be integers such that

$$(x - y)^2 + (y - z)^2 + (z - x)^2 = xyz.$$

Prove that $x^3 + y^3 + z^3$ is divisible by $x + y + z + 6$.

Problem 1.6 Let a, b, c be distinct real numbers. Prove that the following equality cannot hold:

$$\sqrt[3]{a - b} + \sqrt[3]{b - c} + \sqrt[3]{c - a} = 0.$$

Problem 1.7 Prove that the number

$$\sqrt[3]{45 + 29\sqrt{2}} + \sqrt[3]{45 - 29\sqrt{2}}$$

is a rational number.

Problem 1.8 Let a, b, c be rational numbers such that

$$a + b\sqrt[3]{2} + c\sqrt[3]{4} = 0.$$

Prove that $a = b = c = 0$.

Problem 1.9 Let r be a real number such that

$$\sqrt[3]{r} + \frac{1}{\sqrt[3]{r}} = 3.$$

Determine the value of

$$r^3 + \frac{1}{r^3}.$$

Problem 1.10 Find the locus of points (x, y) for which

$$x^3 + y^3 + 3xy = 1.$$

Problem 1.11 Let n be a positive integer. Prove that the number

$$3^{3^n}\left(3^{3^n} + 1\right) + 3^{3^n+1} - 1.$$

is not a prime.

Problem 1.12 Let S be the set of integers x such that $x = a^3 + b^3 + c^3 - 3abc$, for some integers a, b, c. Prove that if $x, y \in S$, then $xy \in S$.

Problem 1.13 Let a, b, c be distinct positive integers and let k be a positive integer such that

$$ab + bc + ca \geq 3k^2 - 1.$$

Prove that

$$\frac{a^3 + b^3 + c^3}{3} - abc \geq 3k.$$

Problem 1.14 Let a, b, c be the side lengths of a triangle. Prove that

$$\sqrt[3]{\frac{a^3 + b^3 + c^3 + 3abc}{2}} \geq \max(a, b, c).$$

Problem 1.15 Find the least real number r such that for each triangle with side lengths a, b, c,

$$\frac{\max(a, b, c)}{\sqrt[3]{a^3 + b^3 + c^3 + 3abc}} < r.$$

Problem 1.16 Find all integers that can be represented as $a^3 + b^3 + c^3 - 3abc$ for some positive integers $a, b,$ and c.

Problem 1.17 Find all pairs (x, y) of integers such that

$$xy + \frac{x^3 + y^3}{3} = 2007.$$

Problem 1.18 Let k be an integer and let

$$n = \sqrt[3]{k + \sqrt{k^2 - 1}} + \sqrt[3]{k - \sqrt{k^2 - 1}} + 1.$$

Prove that $n^3 - 3n^2$ is an integer.

1.2 Cauchy–Schwarz Revisited

Let $a_1, a_2, \ldots, a_n, b_1, b_2, \ldots, b_n$ be nonzero real numbers. Then

$$\left(a_1^2 + a_2^2 + \cdots + a_n^2\right)\left(b_1^2 + b_2^2 + \cdots + b_n^2\right) \geq (a_1 b_1 + a_2 b_2 + \cdots + a_n b_n)^2 \quad (*)$$

with equality if and only if

$$\frac{a_1}{b_1} = \frac{a_2}{b_2} = \cdots = \frac{a_n}{b_n}.$$

This is the well-known Cauchy–Schwarz inequality. The standard elementary proof uses the properties of the quadratic function: consider the function

$$f(x) = (a_1 - b_1 x)^2 + (a_2 - b_2 x)^2 + \cdots + (a_n - b_n x)^2.$$

Clearly, $f(x) \geq 0$ for all real x, therefore, being a quadratic function, its discriminant Δ must be negative or zero. The inequality follows by observing that

$$\Delta = 4(a_1 b_1 + a_2 b_2 + \cdots + a_n b_n)^2 - 4(a_1^2 + a_2^2 + \cdots + a_n^2)(b_1^2 + b_2^2 + \cdots + b_n^2).$$

If we have equality in $(*)$, then $\Delta = 0$ and the equation $f(x) = 0$ has a real root x_0. But then

$$(a_1 - b_1 x_0)^2 + (a_2 - b_2 x_0)^2 + \cdots + (a_n - b_n x_0)^2 = f(x_0) = 0,$$

so that $a_1 - b_1 x_0 = a_2 - b_2 x_0 = \cdots = a_n - b_n x_0 = 0$ and

$$\frac{a_1}{b_1} = \frac{a_2}{b_2} = \cdots = \frac{a_n}{b_n} = x_0.$$

Conversely, if

$$\frac{a_1}{b_1} = \frac{a_2}{b_2} = \cdots = \frac{a_n}{b_n}$$

then the equation $f(x) = 0$ has a real root, so $\Delta \geq 0$. Since Δ cannot be positive, it follows that $\Delta = 0$ and we have equality in $(*)$.

Another proof uses a simple lemma which can also be helpful in proving a large number of algebraic inequalities:

If a, b, x, y are real numbers and $x, y > 0$, then the following inequality holds:

$$\frac{a^2}{x} + \frac{b^2}{y} \geq \frac{(a+b)^2}{x+y}.$$

The proof is straightforward. Clearing out denominators yields

$$a^2 y(x + y) + b^2 x(x + y) \geq (a + b)^2 xy,$$

which readily simplifies to the obvious $(ay - bx)^2 \geq 0$. We see that the equality holds if and only if $ay = bx$, that is, if

$$\frac{a}{x} = \frac{b}{y}.$$

Applying the lemma twice, we can extend the inequality to three pairs of numbers. Indeed,

$$\frac{a^2}{x} + \frac{b^2}{y} + \frac{c^2}{z} \geq \frac{(a+b)^2}{x+y} + \frac{c^2}{z} \geq \frac{(a+b+c)^2}{x+y+z},$$

and a simple inductive argument shows that

$$\frac{a_1^2}{x_1} + \frac{a_2^2}{x_2} + \cdots + \frac{a_n^2}{x_n} \geq \frac{(a_1 + a_2 + \cdots + a_n)^2}{x_1 + x_2 + \cdots + x_n}$$

for all real numbers $a_1, a_2, \ldots, a_n$ and $x_1, x_2, \ldots, x_n > 0$, with equality if and only if

$$\frac{a_1}{x_1} = \frac{a_2}{x_2} = \cdots = \frac{a_n}{x_n}.$$

Returning to Cauchy–Schwarz, let us use the lemma in the general case:

$$a_1^2 + a_2^2 + \cdots + a_n^2 = \frac{a_1^2 b_1^2}{b_1^2} + \frac{a_2^2 b_2^2}{b_2^2} + \cdots + \frac{a_n^2 b_n^2}{b_n^2} \geq \frac{(a_1 b_1 + a_2 b_2 + \cdots + a_n b_n)^2}{b_1^2 + b_2^2 + \cdots + b_n^2}.$$

This yields

$$\left(a_1^2 + a_2^2 + \cdots + a_n^2\right)\left(b_1^2 + b_2^2 + \cdots + b_n^2\right) \geq (a_1 b_1 + a_2 b_2 + \cdots + a_n b_n)^2$$

and the equality holds if and only if

$$\frac{a_1}{b_1} = \frac{a_2}{b_2} = \cdots = \frac{a_n}{b_n}.$$

Let us see our lemma at work!

Problem 1.19 Let a, b, c be positive real numbers. Prove that

$$\frac{a}{b+c} + \frac{b}{c+a} + \frac{c}{a+b} \geq \frac{3}{2}.$$

Solution Observe that

$$\frac{a}{b+c} + \frac{b}{c+a} + \frac{c}{a+b} = \frac{a^2}{ab+ac} + \frac{b^2}{bc+ba} + \frac{c^2}{ca+cb} \geq \frac{(a+b+c)^2}{2(ab+bc+ca)},$$

so it suffices to prove

$$\frac{(a+b+c)^2}{2(ab+bc+ca)} \geq \frac{3}{2}.$$

A short computation shows that this is equivalent to

$$a^2 + b^2 + c^2 \geq ab + bc + ca,$$

which yields

$$(a-b)^2 + (b-c)^2 + (c-a)^2 \geq 0.$$

Problem 1.20 Let a and b be positive real numbers. Prove that

$$8\left(a^4 + b^4\right) \geq (a+b)^4.$$

Solution We apply the lemma twice:

$$a^4 + b^4 = \frac{a^4}{1} + \frac{b^4}{1} \geq \frac{(a^2+b^2)^2}{2} \geq \frac{(\frac{(a+b)^2}{2})^2}{2} = \frac{(a+b)^4}{8}.$$

Try to use the lemma to prove the following inequalities.

Problem 1.21 Let $x, y, z > 0$. Prove that

$$\frac{2}{x+y} + \frac{2}{y+z} + \frac{2}{z+x} \geq \frac{9}{x+y+z}.$$

Problem 1.22 Let a, b, x, y, z be positive real numbers. Prove that

$$\frac{x}{ay+bz} + \frac{y}{az+bx} + \frac{z}{ax+by} \geq \frac{3}{a+b}.$$

Problem 1.23 Let $a, b, c > 0$. Prove that

$$\frac{a^2+b^2}{a+b} + \frac{b^2+c^2}{b+c} + \frac{a^2+c^2}{a+c} \geq a+b+c.$$

Problem 1.24 Let a, b, c be positive numbers such that $abc = 1$. Prove that

$$\frac{1}{a^3(b+c)} + \frac{1}{b^3(a+c)} + \frac{1}{c^3(a+b)} \geq \frac{3}{2}.$$

Problem 1.25 Let $x, y, z > 0$. Prove that

$$\frac{x}{x+2y+3z} + \frac{y}{y+2z+3x} + \frac{z}{z+2x+3y} \geq \frac{1}{2}.$$

Problem 1.26 Let $x, y, z > 0$. Prove that

$$\frac{x^2}{(x+y)(x+z)} + \frac{y^2}{(y+z)(y+x)} + \frac{z^2}{(z+x)(z+y)} \geq \frac{3}{4}.$$

Problem 1.27 Let a, b, c, d, e be positive real numbers. Prove that

$$\frac{a}{b+c} + \frac{b}{c+d} + \frac{c}{d+e} + \frac{d}{e+a} + \frac{e}{a+b} \geq \frac{5}{2}.$$

Problem 1.28 Let a, b, c be positive real numbers such that

$$ab+bc+ca = \frac{1}{3}.$$

Prove that

$$\frac{a}{a^2-bc+1} + \frac{b}{b^2-ca+1} + \frac{c}{c^2-ab+1} \geq \frac{1}{a+b+c}.$$

Problem 1.29 Let a, b, c be positive real numbers such that $abc = 1$. Prove that

$$\frac{a+b+1}{a+b^2+c^3} + \frac{b+c+1}{b+c^2+a^3} + \frac{c+a+1}{c+a^2+b^3} \le \frac{(a+1)(b+1)(c+1)+1}{a+b+c}.$$

Problem 1.30 Let a and b be positive real numbers. Prove that

$$\frac{a^3+b^3}{a^4+b^4} \cdot \frac{a+b}{a^2+b^2} \ge \frac{a^4+b^4}{a^6+b^6}.$$

Problem 1.31 Let a, b, c be positive real numbers such that $ab+bc+ca \ge 3$. Prove that

$$\frac{a}{\sqrt{a+b}} + \frac{b}{\sqrt{b+c}} + \frac{c}{\sqrt{c+a}} \ge \frac{3}{\sqrt{2}}.$$

Problem 1.32 Let a, b, c be positive real numbers. Prove that

$$\frac{a}{b(b+c)^2} + \frac{b}{c(c+a)^2} + \frac{c}{a(a+b)^2} \ge \frac{9}{4(ab+bc+ca)}.$$

Problem 1.33 Let a, b, c be positive real numbers such that

$$\frac{1}{a^2+b^2+1} + \frac{1}{b^2+c^2+1} + \frac{1}{c^2+a^2+1} \ge 1.$$

Prove that

$$ab + bc + ca \le 3.$$

1.3 Easy Ways Through Absolute Values

Everybody knows that sometimes solving equations or inequalities with absolute values can be boring. Most of the students facing such problems begin by writing the absolute values in an explicit manner. Let us consider, for instance, the following simple equation:

Problem 1.34 Solve the equation $|2x - 1| = |x + 3|$.

Solution We have

$$|2x - 1| = \begin{cases} -2x + 1, & x \le \frac{1}{2}, \\ 2x - 1, & x > \frac{1}{2}, \end{cases} \quad \text{and} \quad |x + 3| = \begin{cases} -x - 3, & x \le -3, \\ x + 3, & x > -3. \end{cases}$$

If $x \le -3$, the equation becomes $-2x + 1 = -x - 3$; hence $x = 4$. But $4 > -3$, so we have no solutions in this case. If $-3 < x \le \frac{1}{2}$, we obtain $-2x + 1 = x + 3$, so

$x = -\frac{2}{3} \in (-3, \frac{1}{2}]$. Finally, if $x > \frac{1}{2}$, we obtain $x = 4$ again. We conclude that the solutions are $x = -\frac{2}{3}$ and $x = 4$.

Nevertheless, there is an easier way to solve the equation by noticing that $|a| = |b|$ if and only if $a = \pm b$, so that it is not necessary to write the absolute values in an explicit form. Here are some properties of the absolute values that might be useful in solving equations and inequalities:

$$|ab| = |a||b|,$$

$$\left|\frac{a}{b}\right| = \frac{|a|}{|b|},$$

$$|a + b| \le |a| + |b|,$$

with equality if and only if $ab \ge 0$,

$$|a - b| \le |a| + |b|,$$

with equality if and only if $ab \le 0$.

The last two inequalities can be written in a general form:

$$|\pm a_1 \pm a_2 \pm \cdots \pm a_n| \le |a_1| + |a_2| + \cdots + |a_n|.$$

Problem 1.35 Solve the equation $|x - 1| + |x - 4| = 2$.

Solution Observe that

$$|x - 1| + |x - 4| \ge |(x - 1) - (x - 4)| = |3| = 3 > 2,$$

hence the equation has no solutions.

Problem 1.36 Solve the equation $|x - 1| + |x| + |x + 1| = x + 2$.

Solution We have

$$x + 2 = |x - 1| + |x| + |x + 1| \ge |(x - 1) - (x + 1)| + |x| = |x| + 2.$$

On the other hand, clearly $x \le |x|$, thus all inequalities must be equalities: that is $x + 2 = |x| + 2$ and

$$|x - 1| + |x| + |x + 1| = |(x - 1) - (x + 1)| + |x|.$$

This implies that $x \ge 0$ and the expressions $x - 1$ and $x + 1$ are of different sign, so $x \in [-1, 1]$. Finally, the solution is $x \in [0, 1]$.

Problem 1.37 Find the minimum value of the expression:

$$E(x) = |x - 1| + |x - 2| + \cdots + |x - 100|,$$

where x is a real number.

Solution Observe that for $k = 1, 2, \ldots, 50$, we have

$$|x - k| + |x - (101 - k)| \geq 101 - 2k,$$

with equality if $x \in [k, 101 - k]$. Adding these inequalities yields $E(x) \geq 2500$, and the equality holds for all x such that

$$x \in \bigcap_{1 \leq k \leq 50} [k, 101 - k] = [50, 51].$$

Problem 1.38 Find the values of a for which the equation

$$(x - 1)^2 = |x - a|$$

has exactly three solutions.

Solution Observe that $(x - 1)^2 = |x - a|$ if and only if

$$x - a = \pm (x - 1)^2,$$

that is, if and only if

$$a = x \pm (x - 1)^2.$$

The number of solutions of the equation is equal to the number of intersection points between the line $y = a$ and the graphs of the functions

$$f(x) = x + (x - 1)^2 = x^2 - x + 1$$

and

$$g(x) = x - (x - 1)^2 = -x^2 + 3x - 1.$$

The graph of f is a parabola with vertex $B(\frac{1}{2}, \frac{3}{4})$ and the graph of g a parabola with vertex $C(\frac{3}{2}, \frac{5}{4})$. Now, since the equation $f(x) = g(x)$ is a quadratic with one real root, it follows that the graphs are tangent to each other at point $A(1, 1)$. We deduce that the line $y = a$ intersects the two graphs at three points if and only is it passes through one of the points A, B, C; that is, if $a \in \{\frac{3}{4}, 1, \frac{5}{4}\}$.

Try to use some of the ideas above in solving the following problems:

Problem 1.39 Solve the equation $|x - 3| + |x + 1| = 4$.

Problem 1.40 Show that the equation $|2x - 3| + |x + 1| + |5 - x| = 0.99$ has no solutions.

Problem 1.41 Let $a, b > 0$. Find the values of m for which the equation

$$|x - a| + |x - b| + |x + a| + |x + b| = m(a + b)$$

has at least one real solution.

Problem 1.42 Find all possible values of the expression

$$E(x, y, z) = \frac{|x + y|}{|x| + |y|} + \frac{|y + z|}{|y| + |z|} + \frac{|z + x|}{|z| + |x|},$$

where x, y, z are nonzero real numbers.

Problem 1.43 Find all positive real numbers $x, x_1, x_2, \ldots, x_n$ such that

$$\left| \log(xx_1) \right| + \left| \log(xx_2) \right| + \cdots + \left| \log(xx_n) \right|$$

$$+ \left| \log\left(\frac{x}{x_1}\right) \right| + \left| \log\left(\frac{x}{x_2}\right) \right| + \cdots + \left| \log\left(\frac{x}{x_n}\right) \right|$$

$$= \left| \log x_1 + \log x_2 + \cdots + \log x_n \right|.$$

Problem 1.44 Prove that for all real numbers a, b, we have

$$\frac{|a + b|}{1 + |a + b|} \leq \frac{|a|}{1 + |a|} + \frac{|b|}{1 + |b|}.$$

Problem 1.45 Let n be an odd positive integer and let $x_1, x_2, \ldots, x_n$ be distinct real numbers. Find all one-to-one functions

$$f : \{x_1, x_2, \ldots, x_n\} \to \{x_1, x_2, \ldots, x_n\}$$

such that

$$\left| f(x_1) - x_1 \right| = \left| f(x_2) - x_2 \right| = \cdots = \left| f(x_n) - x_n \right|.$$

Problem 1.46 Suppose the sequence $a_1, a_2, \ldots, a_n$ satisfies the following conditions:

$$a_1 = 0, \qquad |a_2| = |a_1 + 1|, \ldots, \qquad |a_n| = |a_{n-1} + 1|.$$

Prove that

$$\frac{a_1 + a_2 + \cdots + a_n}{n} \geq -\frac{1}{2}.$$

Problem 1.47 Find real numbers a, b, c such that

$$|ax + by + cz| + |bx + cy + az| + |cx + ay + bz| = |x| + |y| + |z|,$$

for all real numbers x, y, z.

1.4 Parameters

We start with the following problem.

Problem 1.48 Solve the equation:

$$x^3(x+1) = 2(x+a)(x+2a),$$

where a is a real parameter.

Solution The equation is equivalent to

$$x^4 + x^3 - 2x^2 - 6ax - 4a^2 = 0.$$

This fourth degree equation is difficult to solve. We might try to factor the left-hand side, but without some appropriate software, the process would get quite complicated. What if we think of a as the unknown and x as the parameter? In this case, the equation can be written as a quadratic:

$$4a^2 + 6xa - x^4 - x^3 + 2x^2 = 0,$$

whose discriminant is

$$36x^2 + 16(x^4 + x^3 - 2x^2) = 4x^2(2x+1)^2,$$

fortunately a square. Solving for a, we obtain the solutions $a_1 = -\frac{1}{2}x^2 - x$ and $a_2 = \frac{1}{2}x^2 - \frac{1}{2}x$, yielding the factorization

$$4a^2 + 6ax - x^4 - x^3 + 2x^2 = 4\left(a + \frac{1}{2}x^2 + x\right)\left(a - \frac{1}{2}x^2 + \frac{1}{2}x\right)$$

$$= -(x^2 + 2x + 2a)(x^2 - x - 2a).$$

Finally, we obtain the solutions $x_{1,2} = -1 \pm \sqrt{1 - 2a}$, $x_{3,4} = \frac{1}{2} \pm \frac{1}{2}\sqrt{1 + 8a}$.

Problem 1.49 Solve the equation $\sqrt{5 - x} = 5 - x^2$.

Solution There is no parameter! Note, however, that we must have $x \in [-\sqrt{5}, \sqrt{5}]$, since the left-hand side is nonnegative. Squaring both sides, we obtain the equation

$$x^4 - 10x^2 + x + 20 = 0$$

and, if we are lucky, we might observe the factorization

$$(x^2 + x - 5)(x^2 - x - 4) = 0.$$

What if we are not lucky? Well, let us introduce a parameter ourselves: replace 5 by a, where $a > 0$: $\sqrt{a - x} = a - x^2$. Squaring both sides yields the equation

$$x^4 - 2ax^2 + x + a^2 - a = 0.$$

Consider this as a quadratic in a with x as a parameter:

$$a^2 - (2x^2 + 1)a + x^4 + x = 0.$$

The discriminant of the quadratic is

$$\Delta = (2x - 1)^2$$

and thus the roots are $a_1 = x^2 + x$ and $a_2 = x^2 - x + 1$. It follows that

$$a^2 - (2x^2 + 1)a + x^4 + x = (a - x^2 - x)(a - x^2 + x - 1).$$

Returning to $a = 5$, we arrive at the desired factorization.

The equations $x^2 + x - 5 = 0$ and $x^2 - x - 4 = 0$ have the solutions $x_{1,2} = \frac{1}{2}(-1 \pm \sqrt{21})$ and $x_{3,4} = \frac{1}{2}(1 \pm \sqrt{17})$, respectively. Only two of them belong to the interval $[-\sqrt{5}, \sqrt{5}]$, and therefore the solutions of the initial equation are $\frac{1}{2}(-1 + \sqrt{21})$ and $\frac{1}{2}(1 - \sqrt{17})$.

Here are some suggested problems.

Problem 1.50 Solve the equation

$$x = \sqrt{a - \sqrt{a + x}},$$

where $a > 0$ is a parameter.

Problem 1.51 Let a be a nonzero real number. Solve the equation

$$a^3 x^4 + 2a^2 x^2 + x + a + 1 = 0.$$

Problem 1.52 Let $a \in (0, \frac{1}{4})$. Solve the equation

$$x^2 + 2ax + \frac{1}{16} = -a + \sqrt{a^2 + x - \frac{1}{16}}.$$

Problem 1.53 Find the positive solutions of the following system of equations:

$$\begin{cases} \frac{a^2}{x^2} - \frac{b^2}{y^2} = 8(y^4 - x^4), \\ ax - by = x^4 - y^4 \end{cases}$$

where $a, b > 0$ are parameters.

Problem 1.54 Let $a, b, c > 0$. Solve the system of equations

$$\begin{cases} ax - by + \frac{1}{xy} = c, \\ bz - cx + \frac{1}{zx} = a, \\ cy - az + \frac{1}{yz} = b. \end{cases}$$

Problem 1.55 Solve the equation

$$x + a^3 = \sqrt[3]{a - x},$$

where a is a real parameter.

1.5 Take the Conjugate!

Let a and b be positive numbers. Recall that the $AM - GM$ states that $\frac{a+b}{2} \geq \sqrt{ab}$.
Can we estimate the difference between the arithmetic and geometric means?

Problem 1.56 Prove that if $a \geq b > 0$, then

$$\frac{(a-b)^2}{8a} \leq \frac{a+b}{2} - \sqrt{ab} \leq \frac{(a-b)^2}{8b}.$$

Solution By taking the conjugate of the difference above, we obtain

$$\frac{a+b}{2} - \sqrt{ab} = \frac{(\frac{a+b}{2})^2 - ab}{\frac{a+b}{2} + \sqrt{ab}} = \frac{(a-b)^2}{2(a+b+2\sqrt{ab})}.$$

Now, the conclusion follows by observing that $b \leq \sqrt{ab} \leq a$, and that

$$8b \leq 2(a+b+2\sqrt{ab}) \leq 8a.$$

Problem 1.57 Evaluate the integer part of the number

$$A = \frac{1}{\sqrt{2}} + \frac{1}{\sqrt{3}} + \cdots + \frac{1}{\sqrt{10000}}.$$

Solution Observe that

$$\frac{1}{\sqrt{k}} = \frac{2}{\sqrt{k} + \sqrt{k}} < \frac{2}{\sqrt{k} + \sqrt{k-1}}.$$

Taking the conjugate, we obtain

$$\frac{1}{\sqrt{k}} < 2(\sqrt{k} - \sqrt{k-1}).$$

Setting $k = 2, 3, \ldots, 10000$ and adding up yields

$$A < 2(\sqrt{10000} - 1) = 198.$$

Similarly,

$$\frac{1}{\sqrt{k}} = \frac{2}{\sqrt{k} + \sqrt{k}} > \frac{2}{\sqrt{k+1} + \sqrt{k}} = 2(\sqrt{k+1} - \sqrt{k}),$$

hence

$$A > 2(\sqrt{10001} - \sqrt{2}) > 197.$$

It follows that the integer part of the number A equals 197.

Problem 1.58 Let n be a positive integer. Prove that $\lfloor (2 + \sqrt{3})^n \rfloor$ is an odd number.

Solution Observe that the number

$$x_n = (2 + \sqrt{3})^n + (2 - \sqrt{3})^n$$

is an even integer. Indeed, $x_1 = 4$, $x_2 = 14$ and

$$x_{n+2} = 4x_{n+1} - x_n,$$

for all $n \geq 1$, so the assertion follows inductively. Since $0 < (2 - \sqrt{3})^n < 1$, we have

$$y_n = \lfloor (2 + \sqrt{3})^n \rfloor = x_n - 1,$$

hence y_n is odd.

Problem 1.59 Solve the equation

$$\sqrt{1 + mx} = x + \sqrt{1 - mx},$$

where m is a real parameter.

Solution We can try squaring both terms, but since it is difficult to control the sign of the right-hand side, we prefer to write the equation as follows:

$$\sqrt{1 + mx} - \sqrt{1 - mx} = x.$$

Taking the conjugate yields the equivalent form

$$\frac{2mx}{\sqrt{1 + mx} + \sqrt{1 - mx}} = x.$$

We obtain $x = 0$ as solution and, for $x \neq 0$:

$$2m = \sqrt{1 + mx} + \sqrt{1 - mx}.$$

Thus m is necessarily positive. We now square and obtain

$$2m^2 - 1 = \sqrt{1 - m^2 x^2}.$$

We obtain another condition for m, that is $2m^2 - 1 \geq 0$, and squaring again we finally obtain the solutions

$$x = \pm 2\sqrt{1 - m^2},$$

where $m \in [\frac{1}{\sqrt{2}}, 1]$.

Take the conjugate in the following problems:

Problem 1.60 Let a and b be distinct positive numbers and let $A = \frac{a+b}{2}$, $B = \sqrt{ab}$. Prove the inequality

$$B < \frac{(a-b)^2}{8(A-B)} < A.$$

Problem 1.61 Let m, n be positive integers with $m < n$. Find a closed form for the sum

$$\frac{1}{\sqrt{m}+\sqrt{m+1}} + \frac{1}{\sqrt{m+1}+\sqrt{m+2}} + \cdots + \frac{1}{\sqrt{n-1}+\sqrt{n}}.$$

Problem 1.62 For any positive integer n, let

$$f(n) = \frac{4n + \sqrt{4n^2-1}}{\sqrt{2n+1}+\sqrt{2n-1}}.$$

Evaluate the sum $f(1) + f(2) + \cdots + f(40)$.

Problem 1.63 Let a and b be distinct real numbers. Solve the equation

$$\sqrt{x-b^2} - \sqrt{x-a^2} = a - b.$$

Problem 1.64 Solve the following equation, where m is a real parameter:

$$\sqrt{x+\sqrt{x}} - \sqrt{x-\sqrt{x}} = m\sqrt{\frac{x}{x+\sqrt{x}}}.$$

Problem 1.65 Prove that for every positive integer k, there exists a positive integer n_k such that

$$\left(\sqrt{3}-\sqrt{2}\right)^k = \sqrt{n_k} - \sqrt{n_k-1}.$$

Problem 1.66 Let a and b be nonzero integers with $|a| \le 100$, $|b| \le 100$. Prove that

$$\left|a\sqrt{2} + b\sqrt{3}\right| \ge \frac{1}{350}.$$

Problem 1.67 Let n be a positive integer. Prove that

$$\left\lfloor \left(\frac{1+\sqrt{5}}{2}\right)^{4n-2} \right\rfloor - 1$$

is a perfect square.

Problem 1.68 Consider the sequence

$$a_n = \sqrt{1 + \left(1 + \frac{1}{n}\right)^2} + \sqrt{1 + \left(1 - \frac{1}{n}\right)^2}, \quad n \geq 1.$$

Prove that

$$\frac{1}{a_1} + \frac{1}{a_2} + \cdots + \frac{1}{a_{20}}$$

is an integer.

Problem 1.69 Prove that

$$\sum_{n=1}^{9999} \frac{1}{(\sqrt{n} + \sqrt{n+1})(\sqrt[4]{n} + \sqrt[4]{n+1})} = 9.$$

Problem 1.70 Consider the sequence

$$a_n = 2 - \frac{1}{n^2 + \sqrt{n^4 + \frac{1}{4}}}, \quad n \geq 1.$$

Prove that

$$\sqrt{a_1} + \sqrt{a_2} + \cdots + \sqrt{a_{119}}$$

is an integer.

1.6 Inequalities with Convex Functions

A real-valued function f defined on an interval $I \subset \mathbf{R}$ is called *convex* if for all $x_A, x_B \in I$ and for any $\lambda \in [0, 1]$ the following inequality holds:

$$f\left(\lambda x_A + (1 - \lambda)x_B\right) \leq \lambda f(x_A) + (1 - \lambda)f(x_B).$$

Although the definition seems complicated, it has a very simple geometrical interpretation. Let us assume that f is a continuous function. Then f is convex on I if and only if no matter how we choose two points on the function's graph, the segment joining these points lies above the graph (see Fig. 1.1).

To see why, let $A(x_A, y_A)$ and $B(x_B, y_B)$ be two points on the graph of f, with $x_A < x_B$, and let $M(x_M, f(x_M))$ be an arbitrary point on the graph with $x_A < x_M < x_B$. If $N(x_M, y_N)$ is on the segment AB, it suffices to verify that $f(x_M) \leq y_N$. If we let

$$\lambda = \frac{x_B - x_M}{x_B - x_A},$$

Fig. 1.1

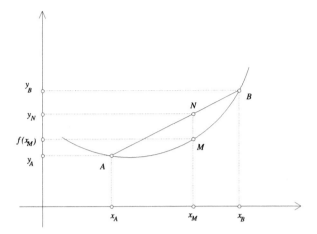

then $\lambda \in [0, 1]$ and

$$x_M = \lambda x_A + (1 - \lambda)x_B.$$

Now it is easy to see that

$$\lambda = \frac{BN}{BA} = \frac{y_B - y_N}{y_B - y_A},$$

that is,

$$y_N = \lambda y_A + (1 - \lambda)y_B = \lambda f(x_A) + (1 - \lambda)f(x_B).$$

Thus the condition $f(x_M) \leq y_N$ is equivalent to the convexity condition

$$f\big(\lambda x_A + (1 - \lambda)x_B\big) \leq \lambda f(x_A) + (1 - \lambda)f(x_B).$$

Observe that unless f is linear, the equality holds if and only if $x_A = x_B$. Hence, if f is not linear and $x_A \neq x_B$, then the inequality is a strict one.

It can be shown that if f is a convex function on the interval I then for any $x_1, x_2, \ldots, x_n \in I$ and any positive numbers $\lambda_1, \lambda_2, \ldots, \lambda_n$ such that $\lambda_1 + \lambda_2 + \cdots + \lambda_n = 1$, we have

$$f(\lambda_1 x_1 + \lambda_2 x_2 + \cdots + \lambda_n x_n) \leq \lambda_1 f(x_1) + \lambda_2 f(x_2) + \cdots + \lambda_n f(x_n).$$

A real-valued function f defined on an interval $I \subset \mathbf{R}$ is called *concave* if for all $x_A, x_B \in I$ and for any $\lambda \in [0, 1]$ the following inequality holds:

$$f\big(\lambda x_A + (1 - \lambda)x_B\big) \geq \lambda f(x_A) + (1 - \lambda)f(x_B).$$

Similar inequalities are valid for concave functions, replacing everywhere $\leq$ by $\geq$.

It is known that if f is twice differentiable, then f is convex on I if an only if $f'' \geq 0$ on I (and concave if $f'' \leq 0$ on I).

Problem 1.71 Let f be a convex function on the interval $I \subset R$. Prove that for any $a < b < c$ in I, we have

$$f(a - b + c) \leq f(a) - f(b) + f(c).$$

Solution Since $b \in (a, c)$, there exists λ such that $b = \lambda a + (1 - \lambda)c$ (take $\lambda = \frac{c-b}{c-a}$ and notice that $\lambda \in [0, 1]$). Since f is convex, we have

$$f(b) \leq \lambda f(a) + (1 - \lambda)f(c).$$

On the other hand, we can see that

$$a - b + c = a - \left(\lambda a + (1 - \lambda)c\right) + c = (1 - \lambda)a + \lambda c.$$

Therefore, by convexity

$$f(a - b + c) \leq (1 - \lambda)f(a) + \lambda f(c).$$

Adding up the two inequalities, we obtain

$$f(a - b + c) + f(b) \leq f(a) + f(c),$$

the desired result.

A simple induction argument shows that if f is convex and

$$a_1 < a_2 < \cdots < a_{2n+1},$$

then

$$f(a_1 - a_2 + a_3 - \cdots - a_{2n} + a_{2n+1}) \leq f(a_1) - f(a_2) + f(a_3) - \cdots - f(a_{2n})$$
$$+ f(a_{2n+1}).$$

If f is concave, then the inequality is reversed.

Problem 1.72 Let $a, b, c > 0$. Prove the inequality

$$\frac{a}{b+c} + \frac{b}{a+c} + \frac{c}{a+b} \geq \frac{3}{2}.$$

Solution This is a classical inequality. Usual solutions use Cauchy–Schwarz inequality or simple algebraic inequalities obtained by denoting $b + c = x$, etc. Another proof is given in the section "Cauchy–Schwarz Revisited" of this book.

Set $s = a + b + c$. Then the inequality becomes

$$\frac{a}{s-a} + \frac{b}{s-b} + \frac{c}{s-c} \geq \frac{3}{2}.$$

Consider the function $f(x) = \frac{x}{s-x}$, defined on $(0, s)$. We have

$$f''(x) = \frac{2s}{(s-x)^3} > 0,$$

so that f is convex on $(0, s)$. We then have

$$f\left(\frac{a+b+c}{3}\right) \leq \frac{f(a)+f(b)+f(c)}{3},$$

which means that

$$\frac{a}{s-a} + \frac{b}{s-b} + \frac{c}{s-c} \geq 3f\left(\frac{s}{3}\right) = 3\frac{\frac{s}{3}}{s-\frac{s}{3}} = \frac{3}{2},$$

as desired.

Problem 1.73 Let $f : [1, 13] \to R$ be a convex and integrable function. Prove that

$$\int_1^3 f(x)\,dx + \int_{11}^{13} f(x)\,dx \geq \int_5^9 f(x)\,dx.$$

Solution We have seen that if $a < b < c$

$$f(a - b + c) + f(b) \leq f(a) + f(c).$$

Let $c = a + 10$ and $b = a + 4$. Then

$$f(a + 6) + f(a + 4) \leq f(a) + f(a + 10).$$

If we integrate both sides for $a \in [1, 3]$, and notice that

$$\int_1^3 f(a+6)\,da = \int_7^9 f(x)\,dx, \qquad \int_1^3 f(a+4)\,da = \int_5^7 f(x)\,dx,$$

and

$$\int_1^3 f(a+10)\,da = \int_{11}^{13} f(x)\,dx,$$

the result follows.

Proposed problems:

Problem 1.74 Let $a, b > 0$ and let n be a positive integer. Prove the inequality

$$\frac{a^n + b^n}{2} \geq \left(\frac{a+b}{2}\right)^n.$$

Problem 1.75 Prove that

$$\sqrt[3]{3 - \sqrt[3]{3}} + \sqrt[3]{3 + \sqrt[3]{3}} < 2\sqrt[3]{3}.$$

Problem 1.76 Prove the $AM - GM$ inequality

$$\frac{x_1 + x_2 + \cdots + x_n}{n} \geq \sqrt[n]{x_1 x_2 \cdots x_n},$$

for all $x_1, x_2, \ldots, x_n > 0$.

Problem 1.77 Let $a_1 < a_2 < \cdots < a_{2n+1}$ be positive real numbers. Prove the inequality

$$\sqrt[n]{a_1 - a_2 + a_3 - \cdots - a_{2n} + a_{2n+1}} \geq \sqrt[n]{a_1} - \sqrt[n]{a_2} + \cdots + \sqrt[n]{a_{2n+1}}.$$

Problem 1.78 Let $x, y, z > 0$. Prove that

$$\frac{x}{2x + y + z} + \frac{y}{x + 2y + z} + \frac{z}{x + y + 2z} \leq \frac{3}{4}.$$

Problem 1.79 Prove that if $a, b, c, d > 0$ and $a \leq 1, a + b \leq 5, a + b + c \leq 14$, $a + b + c + d \leq 30$, then

$$\sqrt{a} + \sqrt{b} + \sqrt{c} + \sqrt{d} \leq 10.$$

1.7 Induction at Work

A large number of identities, inequalities etc. can be proved by induction. Sometimes induction is helpful in less obvious situations. We will examine several examples.

Problem 1.80 Let n be a positive integer. Find the roots of the polynomial

$$P_n(X) = 1 + \frac{X}{1!} + \frac{X(X + 1)}{2!} + \cdots + \frac{X(X + 1) \cdots (X + n - 1)}{n!}.$$

Solution For $n = 1$, the polynomial $P_1(X) = 1 + X$ has the root -1. For $n = 2$, $P_2(X) = 1 + \frac{X}{1!} + \frac{X(X+1)}{2!} = \frac{1}{2}(X + 2)(X + 1)$ has roots -1 and -2. It is natural to presume that the roots of P_n are $-1, -2, \ldots, -n$. We prove this assertion by induction on n. Suppose it holds true for n. Then P_n factors as

$$P_n(X) = c(X + 1)(X + 2) \cdots (X + n),$$

the number c being the coefficient of X^n. Checking again the definition of P_n, we see that $c = \frac{1}{n!}$, hence

$$P_n(X) = \frac{1}{n!}(X+1)(X+2)\cdots(X+n).$$

Now, observe that

$$
\begin{aligned}
P_{n+1}(X) &= P_n(X) + \frac{X(X+1)\cdots(X+n)}{(n+1)!}\\
&= \frac{1}{n!}(X+1)(X+2)\cdots(X+n) + \frac{X(X+1)\cdots(X+n)}{(n+1)!}\\
&= \frac{1}{(n+1)!}(X+1)(X+2)\cdots(X+n)(n+1+X),
\end{aligned}
$$

hence the roots of P_{n+1} are $-1, -2, \ldots, -n$ and $-(n+1)$.

Problem 1.81 Let n be a positive integer. Prove the inequality

$$\left(1+\frac{1}{1^3}\right)\left(1+\frac{1}{2^3}\right)\left(1+\frac{1}{3^3}\right)\cdots\left(1+\frac{1}{n^3}\right) < 3.$$

Solution Apparently induction does not work here, since when passing from n to $n+1$, the left-hand side increases while the right-hand side is constant. We can make induction work by proving a stronger result:

$$\left(1+\frac{1}{1^3}\right)\left(1+\frac{1}{2^3}\right)\left(1+\frac{1}{3^3}\right)\cdots\left(1+\frac{1}{n^3}\right) \le 3 - \frac{1}{n}.$$

The assertion is clearly true for $n = 1$. Suppose is true for n and multiply the above inequality by $1 + \frac{1}{(n+1)^3}$. It suffices to prove that

$$\left(3-\frac{1}{n}\right)\left(1+\frac{1}{(n+1)^3}\right) \le 3 - \frac{1}{n+1}.$$

The difference

$$\left(3-\frac{1}{n}\right)\left(1+\frac{1}{(n+1)^3}\right) - 3 + \frac{1}{n+1}$$

factors as

$$-\frac{n^2-n+2}{n(n+1)^3},$$

hence it is negative. The claim is proved.

Problem 1.82 Let p be a prime number. Prove that for any positive integer a the number $a^p - a$ is divisible by p (Fermat's little theorem).

Solution We proceed by induction on a. The statement is true for $a = 1$, so suppose it holds true for some $a > 1$. We want to prove that $(a + 1)^p - (a + 1)$ is divisible by p. The binomial theorem gives

$$(a + 1)^p = \sum_{k=0}^{p} \binom{p}{k} a^{p-k},$$

hence

$$(a + 1)^p - (a + 1) = a^p - a + \sum_{k=1}^{p-1} \binom{p}{k} a^{p-k}.$$

The conclusion follows from the induction hypothesis and from the fact that the binomial coefficients $\binom{p}{k}$ for $k = 1, 2, \ldots, p - 1$ are divisible by p.

Use induction in solving the following problems.

Problem 1.83 Let n be a positive integer. Prove the inequality

$$\left(1 + \frac{1}{2}\right)\left(1 + \frac{1}{2^2}\right)\left(1 + \frac{1}{2^3}\right) \cdots \left(1 + \frac{1}{2^n}\right) < \frac{5}{2}.$$

Problem 1.84 Let n be a positive integer. Prove that the number

$$2^{2^n} - 1$$

has at least n distinct prime divisors.

Problem 1.85 Let a and n be positive integers such that $a < n!$. Prove that a can be represented as a sum of at most n distinct divisors of $n!$.

Problem 1.86 Let $x_1, x_2, \ldots, x_m, y_1, y_2, \ldots, y_n$ be positive integers such that the sums $x_1 + x_2 + \cdots + x_m$ and $y_1 + y_2 + \cdots + y_n$ are equal and less than mn. Prove that in the equality

$$x_1 + x_2 + \cdots + x_m = y_1 + y_2 + \cdots + y_n$$

one can cancel some terms and obtain another equality.

Problem 1.87 The sequence $(x_n)_{n \geq 1}$ is defined by $x_1 = 1$, $x_{2n} = 1 + x_n$ and $x_{2n+1} = \frac{1}{x_{2n}}$ for all $n \geq 1$. Prove that for any positive rational number r there exists an unique n such that $r = x_n$.

Problem 1.88 Let n be a positive integer and let $0 < a_1 < a_2 < \cdots < a_n$ be real numbers. Prove that at least $\binom{n+1}{2}$ of the sums $\pm a_1 \pm a_2 \pm \cdots \pm a_n$ are distinct.

Problem 1.89 Prove that for each positive integer n, there are pairwise relatively prime integers $k_0, k_1, \ldots, k_n$, all strictly greater than 1, such that $k_0 k_1 \cdots k_n - 1$ is the product of two consecutive integers.

Problem 1.90 Prove that for every positive integer n, the number $3^{3^n} + 1$ is the product of at least $2n + 1$ (not necessarily distinct) primes.

Problem 1.91 Prove that for every positive integer n there exists an n-digit number divisible by 5^n all of whose digits are odd.

1.8 Roots and Coefficients

Let $P(X) = a_0 + a_1 X + \cdots + a_n X^n$ be a polynomial and $x_1, x_2, \ldots, x_n$ its roots (real or complex). It is known that the following equalities hold:

$$x_1 + x_2 + \cdots + x_n = -\frac{a_{n-1}}{a_n},$$

$$x_1 x_2 + x_1 x_3 + \cdots + x_{n-1} x_n = \frac{a_{n-2}}{a_n},$$

$$x_1 x_2 x_3 + x_1 x_2 x_4 + \cdots + x_{n-2} x_{n-1} x_n = -\frac{a_{n-3}}{a_n},$$

$$\vdots$$

$$x_1 x_2 \cdots x_n = (-1)^n \frac{a_0}{a_n}.$$

These are usually called Viète's relations.

For instance, for a third degree polynomial

$$P(X) = a_0 + a_1 X + a_2 X^2 + a_3 X^3,$$

we have

$$x_1 + x_2 + x_3 = -\frac{a_2}{a_3},$$

$$x_1 x_2 + x_1 x_3 + x_2 x_3 = \frac{a_1}{a_3},$$

$$x_1 x_2 x_3 = -\frac{a_0}{a_3}.$$

The Viète's relations can be very useful in solving problems not necessarily involving polynomials.

Problem 1.92 Let a, b, c be nonzero real numbers such that

$$(ab + bc + ca)^3 = abc(a + b + c)^3.$$

Prove that a, b, c are terms of a geometric sequence.

Solution Consider the monic polynomial

$$P(X) = X^3 + mX^2 + nX + p,$$

with roots a, b, c. Then, by Viète's relations, we have

$$a + b + c = -m,$$
$$ab + bc + ca = n,$$
$$abc = -p.$$

The given equality yields $n^3 = m^3 p$, hence, if $m \neq 0$, the equation $P(x) = 0$ can be written as

$$x^3 + mx^2 + nx + \frac{n^3}{m^3} = 0,$$

or

$$m^3 x^3 + m^4 x^2 + nm^3 x + n^3 = 0.$$

It is not difficult to factor the left-hand side:

$$(mx + n)(m^2 x^2 - mnx + n^2) + m^3 x(mx + n)$$
$$= (mx + n)(m^2 x^2 + (m^3 - mn)x + n^2).$$

It follows that one of the roots of P is $x_1 = -\frac{n}{m}$ and the other two satisfy the condition $x_2 x_3 = \frac{n^2}{m^2}$ (Viète's relations for the second degree polynomial $m^2 X^2 + (m^3 - mn)X + n^2$). We obtained $x_1^2 = x_2 x_3$, thus the roots are the terms of a geometric sequence. If $m = 0$ then $n = 0$ but in this case, the polynomial $X^3 + p$ cannot have three real roots.

Observation Using appropriate software, one can obtain the factorization

$$(ab + bc + ca)^3 - abc(a + b + c)^3 = (a^2 - bc)(b^2 - ac)(c^2 - ab),$$

and the conclusion follows.

Problem 1.93 Solve in real numbers the system of equations

$$\begin{cases} x + y + z = 4, \\ x^2 + y^2 + z^2 = 14, \\ x^3 + y^3 + z^3 = 34. \end{cases}$$

Solution Consider the monic polynomial

$$P(t) = t^3 + at^2 + bt + c,$$

with roots x, y, z.

Because $x + y + z = 4$, it follows that $a = -4$, hence

$$P(t) = t^3 - 4t^2 + bt + c.$$

We have

$$x^2 + y^2 + z^2 = (x + y + z)^2 - 2(xy + xz + yz).$$

It follows that

$$b = xy + xz + yz = 1.$$

The numbers x, y, z are the roots of P, thus

$$x^3 - 4x^2 + x + c = 0,$$
$$y^3 - 4y^2 + y + c = 0,$$
$$z^3 - 4z^2 + z + c = 0.$$

Adding these equalities and using the equations of the system, we obtain $c = 6$, hence

$$P(t) = t^3 - 4t^2 + t + 6.$$

We observe that $t_1 = -1$ is a root, so P factors as

$$P(t) = (t + 1)(t^2 - 5t + 6),$$

the other two roots being $t_2 = 2$ and $t_3 = 3$. It follows that the solutions of the system are the triple $(-1, 2, 3)$ and all of its permutations.

Problem 1.94 Let a and b be two of the roots of the polynomial $X^4 + X^3 - 1$. Prove that ab is a root of the polynomial $X^6 + X^4 + X^3 - X^2 - 1$.

Solution Let c and d be the other two roots of $X^4 + X^3 - 1$. The Viète's relations yield

$$a + b + c + d = -1,$$
$$ab + ac + ad + bc + bd + cd = 0,$$
$$abc + abd + acd + bcd = 0,$$
$$abcd = -1.$$

Write these equalities in terms of $s = a + b$, $s' = c + d$, $p = ab$ and $p' = cd$ (this is often useful) to obtain

$$s + s' = -1,$$
$$p + p' + ss' = 0,$$

$$ps' + p's = 0,$$
$$pp' = -1.$$

Substituting $p' = -\frac{1}{p}$ and $s' = -1 - s$ in the second and in the third equalities yields

$$p - \frac{1}{p} - s^2 - s = 0$$

and

$$p(-1 - s) - \frac{s}{p} = 0.$$

It follows from the second equality that

$$s = -\frac{p^2}{p^2 + 1}.$$

Plugging this into the first equality gives

$$p - \frac{1}{p} - \frac{p^4}{(p^2 + 1)^2} + \frac{p^2}{p^2 + 1} = 0.$$

A short computation shows that this is equivalent to

$$p^6 + p^4 + p^3 - p^2 - 1 = 0,$$

hence $p = ab$ is a root of the polynomial

$$X^6 + X^4 + X^3 - X^2 - 1.$$

Here are some suggested problems.

Problem 1.95 Let a, b, c be nonzero real numbers such that $a + b + c \neq 0$ and

$$\frac{1}{a} + \frac{1}{b} + \frac{1}{c} = \frac{1}{a + b + c}.$$

Prove that for all odd integers n

$$\frac{1}{a^n} + \frac{1}{b^n} + \frac{1}{c^n} = \frac{1}{a^n + b^n + c^n}.$$

Problem 1.96 Let $a \leq b \leq c$ be real numbers such that

$$a + b + c = 2$$

and

$$ab + bc + ca = 1.$$

Prove that

$$0 \le a \le \frac{1}{3} \le b \le 1 \le c \le \frac{4}{3}.$$

Problem 1.97 Prove that two of the four roots of the polynomial $X^4 + 12X - 5$ add up to 2.

Problem 1.98 Find m and solve the following equation, knowing that its roots form a geometrical sequence:

$$X^4 - 15X^3 + 70X^2 - 120X + m = 0.$$

Problem 1.99 Let $x_1, x_2, \ldots, x_n$ be the roots of the polynomial $X^n + X^{n-1} + \cdots + X + 1$. Prove that

$$\frac{1}{1 - x_1} + \frac{1}{1 - x_2} + \cdots + \frac{1}{1 - x_n} = \frac{n}{2}.$$

Problem 1.100 Let a, b, c be rational numbers and let x_1, x_2, x_3 be the roots of the polynomial $P(X) = X^3 + aX^2 + bX + c$. Prove that if $\frac{x_1}{x_2}$ is a rational number, different from 0 and -1, then x_1, x_2, x_3 are rational numbers.

Problem 1.101 Solve in real numbers the system of equations

$$\begin{cases} x + y + z = 0, \\ x^3 + y^3 + z^3 = 18, \\ x^7 + y^7 + z^7 = 2058. \end{cases}$$

Problem 1.102 Solve in real numbers the system of equations

$$\begin{cases} a + b = 8, \\ ab + c + d = 23, \\ ad + bc = 28, \\ cd = 12. \end{cases}$$

1.9 The Rearrangements Inequality

Let $n \ge 2$ be a positive integer and let $x_1 < x_2 < \cdots < x_n$, $y_1 < y_2 < \cdots < y_n$ be two ordered sequences of real numbers. The rearrangements inequality states that among all the sums of the form

$$S(\sigma) = x_1 y_{\sigma(1)} + x_2 y_{\sigma(2)} + \cdots + x_n y_{\sigma(n)},$$

where σ is a permutation of the numbers $1, 2, \ldots, n$, the maximal one is

$$x_1 y_1 + x_2 y_2 + \cdots + x_n y_n,$$

and the minimal one is

$$x_1 y_n + x_2 y_{n-1} + \cdots + x_n y_1.$$

Indeed, let σ be a permutation for which $S(\sigma)$ is maximal. (Such a permutation exists, since the number of possible sums is finite.) Suppose, by way of contradiction, that one can find i, j, with $1 \leq i < j \leq n$, such that $\sigma(i) > \sigma(j)$. Now, switch $\sigma(i)$ and $\sigma(j)$ to obtain a new permutation σ'. More precisely,

$$\sigma'(k) = \begin{cases} \sigma(k), & \text{for } k \neq i, j, \\ \sigma(j), & \text{for } k = i, \\ \sigma(i), & \text{for } k = j. \end{cases}$$

Observe that

$$S(\sigma') - S(\sigma) = x_i y_{\sigma(j)} + x_j y_{\sigma(i)} - x_i y_{\sigma(i)} - x_j y_{\sigma(j)}$$
$$= (x_i - x_j)(y_{\sigma(j)} - y_{\sigma(i)}) > 0,$$

since $x_i < x_j$ and $y_{\sigma(j)} < y_{\sigma(i)}$. This implies $S(\sigma') > S(\sigma)$, contradicting thus the maximality of $S(\sigma)$.

In a similar manner one can prove that $x_1 y_n + x_2 y_{n-1} + \cdots + x_n y_1$ is the minimal sum.

Observations

1. If we replace the initial conditions with the less restrictive ones

$$x_1 \leq x_2 \leq \cdots \leq x_n, \quad y_1 \leq y_2 \leq \cdots \leq y_n,$$

the conclusion still holds, only in this case there might exist more than one maximal (minimal) sum. Just consider the extreme case $x_1 = x_2 = \cdots = x_n$.
2. If the given sequences have opposite monotonies, then the first sum is minimal and the second one maximal.

The rearrangements inequality has numerous interesting applications. Let us begin with a very simple one.

Problem 1.103 Let a, b, c be real numbers. Prove that

$$a^2 + b^2 + c^2 \geq ab + bc + ca.$$

Solution For symmetry reasons, we can assume that $a \leq b \leq c$. Consider the ordered sequences $x_1 \leq x_2 \leq x_3$ and $y_1 \leq y_2 \leq y_3$, with $x_1 = y_1 = a$, $x_2 = y_2 = b$, and $x_3 = y_3 = c$. The rearrangements inequality then gives

$$x_1 y_1 + x_2 y_2 + x_3 y_3 \geq x_1 y_2 + x_2 y_3 + x_3 y_1,$$

which is exactly what we wanted to prove.

Observations In a similar way we can prove the following more general result. Given the real numbers $x_1, x_2, \ldots, x_n$, we have

$$x_1^2 + x_2^2 + \cdots + x_n^2 \geq x_1 x_{\sigma(1)} + x_2 x_{\sigma(2)} + \cdots + x_n x_{\sigma(n)},$$

where σ is an arbitrary permutation of the numbers $1, 2, \ldots, n$.

Problem 1.104 Let a, b, c be positive real numbers. Prove that

$$a^3 + b^3 + c^3 \geq 3abc.$$

Solution Although this inequality follows directly from the identity we presented in the first chapter of this book, we will give another proof using rearrangements.

Obviously, we can assume with no loss of generality that $a \leq b \leq c$, which also implies $a^2 \leq b^2 \leq c^2$. Therefore,

$$a^3 + b^3 + c^3 = a \cdot a^2 + b \cdot b^2 + c \cdot c^2 \geq a \cdot b^2 + b \cdot c^2 + c \cdot a^2.$$

Now, observe that $a \leq b \leq c$ implies $bc \geq ca \geq ab$, as well. Consequently,

$$\begin{aligned}
3abc &= a \cdot (bc) + b \cdot (ca) + c \cdot (ab) \\
&\leq a \cdot (ca) + b \cdot (ab) + c \cdot (bc) \\
&= a \cdot b^2 + b \cdot c^2 + c \cdot a^2.
\end{aligned}$$

We have thus obtained

$$a^3 + b^3 + c^3 \geq ab^2 + bc^2 + ca^2 \geq 3abc,$$

as desired.

The rearrangements inequality can also be helpful in proving other classical inequalities. Check the following two problems to see alternative proofs of the $AM - GM$ and Cauchy–Schwarz inequalities.

Problem 1.105 Let $a_1, a_2, \ldots, a_n$ be positive real numbers. Prove that

$$\frac{a_1 + a_2 + \cdots + a_n}{n} \geq \sqrt[n]{a_1 a_2 \cdots a_n}.$$

Solution Denote by G the geometric mean of the given numbers and consider the sequences

$$x_1 = \frac{a_1}{G}, \qquad x_2 = \frac{a_1 a_2}{G^2}, \dots, \qquad x_{n-1} = \frac{a_1 a_2 \cdots a_{n-1}}{G^{n-1}}, \qquad x_n = \frac{a_1 a_2 \cdots a_n}{G^n} = 1,$$

$$y_1 = \frac{G}{a_1}, \qquad y_2 = \frac{G^2}{a_1 a_2}, \dots, \qquad y_{n-1} = \frac{G^{n-1}}{a_1 a_2 \cdots a_{n-1}}, \qquad y_n = \frac{G^n}{a_1 a_2 \cdots a_n} = 1.$$

Obviously, if we arrange the x_i's in order:

$$x_{\sigma(1)} \le x_{\sigma(2)} \le \cdots \le x_{\sigma(n)},$$

then

$$y_{\sigma(1)} \ge y_{\sigma(2)} \ge \cdots \ge y_{\sigma(n)},$$

and hence the sum

$$x_{\sigma(1)} y_{\sigma(1)} + x_{\sigma(2)} y_{\sigma(2)} + \cdots + x_{\sigma(n)} y_{\sigma(n)} = n$$

is minimal. But then

$$n \le x_1 y_n + x_2 y_1 + x_3 y_2 + \cdots + x_n y_{n-1} = \frac{a_1}{G} + \frac{a_2}{G} + \cdots + \frac{a_n}{G}.$$

Clearly, this is equivalent to

$$\frac{a_1 + a_2 + \cdots + a_n}{n} \ge G,$$

that is, the $AM - GM$ inequality.

Problem 1.106 Let $a_1, a_2, \dots, a_n$, and $b_1, b_2, \dots, b_n$ be real numbers. Prove that

$$\left(a_1^2 + a_2^2 + \cdots + a_n^2\right)\left(b_1^2 + b_2^2 + \cdots + b_n^2\right) \ge (a_1 b_1 + a_2 b_2 + \cdots + a_n b_n)^2.$$

Solution Set

$$A = \sqrt{a_1^2 + a_2^2 + \cdots + a_n^2}, \qquad B = \sqrt{b_1^2 + b_2^2 + \cdots + b_n^2},$$

and

$$x_1 = \frac{a_1}{A}, \qquad x_2 = \frac{a_2}{A}, \dots, \qquad x_n = \frac{a_n}{A}, \qquad x_{n+1} = \frac{b_1}{B}, \qquad x_{n+2} = \frac{b_2}{B}, \dots,$$

$$x_{2n} = \frac{b_n}{B}.$$

(We discarded the obvious case when all a_i's or all b_i's equal zero.)

We have

$$\sum_{k=1}^{2n} x_k^2 \geq \sum_{k=1}^{2n} x_k x_{\sigma(k)},$$

for any permutation σ. (See the observation at the end of the solution of Problem 1.103.) In particular,

$$2 = \sum_{k=1}^{2n} x_k^2 \geq x_1 x_{n+1} + x_2 x_{n+2} + \cdots + x_n x_{2n} + x_{n+1} x_1 + \cdots + x_{2n} x_n$$

$$= 2 \frac{a_1 b_1 + a_2 b_2 + \cdots + a_n b_n}{AB}.$$

It follows that

$$AB \geq a_1 b_1 + a_2 b_2 + \cdots + a_n b_n.$$

Taking instead

$$x_{n+1} = -\frac{b_1}{B}, \qquad x_{n+2} = -\frac{b_2}{B}, \dots, \qquad x_{2n} = -\frac{b_n}{B},$$

we obtain in a similar way

$$AB \geq -(a_1 b_1 + a_2 b_2 + \cdots + a_n b_n),$$

hence

$$AB \geq |a_1 b_1 + a_2 b_2 + \cdots + a_n b_n|.$$

Squaring the latter gives the Cauchy–Schwarz inequality.

Finally, solving the following old IMO problem is not difficult if one uses rearrangements.

Problem 1.107 Let $a_1, a_2, \ldots, a_n$ be distinct positive integers. Prove that

$$\sum_{k=1}^{n} \frac{a_k}{k^2} \geq \sum_{k=1}^{n} \frac{1}{k}.$$

Solution Rearrange the a_i's in increasing order, as $b_1 < b_2 < \cdots < b_n$. The rearrangements inequality yields

$$\sum_{k=1}^{n} \frac{a_k}{k^2} \geq \sum_{k=1}^{n} \frac{b_k}{k^2},$$

since

$$\frac{1}{1^2} > \frac{1}{2^2} > \cdots > \frac{1}{n^2}.$$

On the other hand, it is not difficult to see that $b_k \geq k$, for all k, hence

$$\sum_{k=1}^{n} \frac{b_k}{k^2} \geq \sum_{k=1}^{n} \frac{k}{k^2} = \sum_{k=1}^{n} \frac{1}{k}.$$

Now, rearrange some terms to solve the following problems.

Problem 1.108 Let a, b, c be positive real numbers. Prove the inequality

$$\frac{a}{b+c} + \frac{b}{c+a} + \frac{c}{a+b} \geq \frac{3}{2}.$$

Problem 1.109 Let a, b, c be positive real numbers. Prove the inequality

$$\frac{a^3}{b^2+c^2} + \frac{b^3}{c^2+a^2} + \frac{c^3}{a^2+b^2} \geq \frac{a+b+c}{2}.$$

Problem 1.110 Let a, b, c be positive real numbers. Prove that

$$a+b+c \leq \frac{a^2+b^2}{2c} + \frac{b^2+c^2}{2a} + \frac{c^2+a^2}{2b} \leq \frac{a^3}{bc} + \frac{b^3}{ca} + \frac{c^3}{ab}.$$

Problem 1.111 Let a, b, c be positive real numbers. Prove the inequality

$$\frac{a^2b(b-c)}{a+b} + \frac{b^2c(c-a)}{b+c} + \frac{c^2a(a-b)}{c+a} \geq 0.$$

Problem 1.112 Let $a_1 \leq a_2 \leq \cdots \leq a_n$ and $b_1 \leq b_2 \leq \cdots \leq b_n$ be two ordered sequences of real numbers. Prove Chebyshev's inequality

$$\frac{a_1+a_2+\cdots+a_n}{n} \cdot \frac{b_1+b_2+\cdots+b_n}{n} \leq \frac{a_1b_1+a_2b_2+\cdots+a_nb_n}{n}.$$

Problem 1.113 Let $a_1 \leq a_2 \leq \cdots \leq a_n$ and $b_1 \leq b_2 \leq \cdots \leq b_n$ be two ordered sequences of positive real numbers. Prove that

$$(a_1+b_1)(a_2+b_2)\cdots(a_n+b_n) \leq (a_1+b_{\sigma(1)})(a_2+b_{\sigma(2)})\cdots(a_n+b_{\sigma(n)}),$$

for any permutation σ.

Chapter 2
Geometry and Trigonometry

2.1 Geometric Inequalities

One of the most basic geometric inequalities is the triangle inequality: in every triangle, the length of one side is less than the sum of the two other sides' lengths. More generally, for any three points A, B, C one has

$$AC + BC \geq AB$$

with equality if and only if C lies on the line segment AB. Many interesting problems can be solved using this simple idea.

Problem 2.1 Let $ABCD$ be a convex quadrilateral and let M, N be the midpoints of AD and BC, respectively. Prove that

$$MN = \frac{AB + CD}{2}$$

if and only if AB is parallel to CD.

Solution Let P be the midpoint of the diagonal AC (see Fig. 2.1). Then MP and PN are parallel to CD and AB, respectively. Moreover, we have $MP = \frac{CD}{2}$ and $PN = \frac{AB}{2}$. Applying the triangle inequality to $\triangle MNP$ gives

$$\frac{AB}{2} + \frac{CD}{2} = PN + MP \geq MN.$$

The equality occurs if and only if P lies on the line segment MN; that is, if the lines MP, PN and MN coincide. Since $MP \parallel CD$ and $NP \parallel AB$, the latter holds true if and only if AB and CD are parallel.

T. Andreescu, B. Enescu, *Mathematical Olympiad Treasures*,
DOI 10.1007/978-0-8176-8253-8_2, © Springer Science+Business Media, LLC 2011

Fig. 2.1

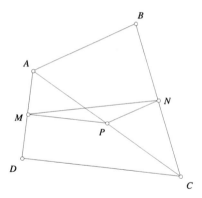

Problem 2.2 Let D be the midpoint of the side BC of triangle ABC. Prove that

$$AD < \frac{AB + AC}{2}.$$

Solution Let the point A' be such that D is the midpoint of AA'. Thus, the quadrilateral $ABA'C$ is a parallelogram and $AD = \frac{AA'}{2}$. In triangle ABA', we have $AA' < AB + BA'$. The conclusion follows observing that $BA' = AC$.

Problem 2.3 Let M be a point inside the triangle ABC. Prove that

$$AB + AC > MB + MC.$$

Solution Let N be the point at which BM intersects AC. Then we successively have

$$AB + AC = AB + AN + NC > BN + NC = BM + MN + NC > BM + MC.$$

Problem 2.4 Let A, B, C and D be four points in space, not in the same plane. Prove that

$$AC \cdot BD < AB \cdot CD + AD \cdot BC.$$

Solution Consider a sphere which passes through the points B, C and D and intersects the segments AB, AC and AD at the points B', C' and D'. The intersection of the sphere and the plane (ABC) is a circle, thus the quadrilateral $BB'C'C$ is cyclic. It follows that triangles ABC and $AC'B'$ are similar; hence

$$\frac{AB'}{AC} = \frac{AC'}{AB} = \frac{B'C'}{BC}.$$

We obtain

$$B'C' = \frac{BC \cdot AB'}{AC}.$$

Analogously, triangles ABD and $AD'B'$ are similar, so that

$$B'D' = \frac{BD \cdot AB'}{AD}.$$

We deduce that

$$\frac{B'C'}{B'D'} = \frac{BC \cdot AD}{AC \cdot BD}.$$

In a similar way, we obtain

$$\frac{C'D'}{B'D'} = \frac{AB \cdot CD}{AC \cdot BD}.$$

But in triangle $A'B'C'$ we have $B'C' + C'D' > B'D'$, and therefore it follows that

$$\frac{BC \cdot AD}{AC \cdot BD} + \frac{AB \cdot CD}{AC \cdot BD} > 1,$$

so that $AC \cdot BD < AB \cdot CD + AD \cdot BC$.

Observation If A, B, C, D are coplanar points, then one can prove that

$$AC \cdot BD \leq AB \cdot CD + AD \cdot BC,$$

with equality if and only if $ABCD$ is a cyclic quadrilateral. The proof is similar, but instead of constructing the sphere, one uses inversion.

Here are some proposed problems.

Problem 2.5 Let $ABCD$ be a convex quadrilateral. Prove that

$$\max(AB + CD, AD + BC) < AC + BD < AB + BC + CD + DA.$$

Problem 2.6 Let M be the midpoint of segment AB. Prove that if O is an arbitrary point, then

$$|OA - OB| \leq 2OM.$$

Problem 2.7 Prove that in an arbitrary triangle, the sum of the lengths of the altitudes is less than the triangle's perimeter.

Problem 2.8 Denote by P the perimeter of triangle ABC. If M is a point in the interior of the triangle, prove that

$$\frac{1}{2}P < MA + MB + MC < P.$$

Problem 2.9 Prove that if A', B' and C' are the midpoints of the sides BC, CA, and AB, respectively, then

$$\frac{3}{4}P < AA' + BB' + CC' < P.$$

Problem 2.10 In the convex quadrilateral $ABCD$, we have

$$AB + BD \leq AC + CD.$$

Prove that $AB < AC$.

Problem 2.11 Consider n red and n blue points in the plane, no three of them being collinear. Prove that one can connect each red point to a blue one with a segment such that no two segments intersect.

Problem 2.12 Let n be an odd positive integer. On some field, n gunmen are placed such that all pairwise distances between them are different. At a signal, every gunman takes out his gun and shoots the closest gunman. Prove that:

(a) at least one gunman survives;
(b) no gunman is shot more than five times;
(c) the trajectories of the bullets do not intersect.

Problem 2.13 Prove that the medians of a given triangle can form a triangle.

Problem 2.14 Let A and B be two points situated on the same side of a line XY. Find the position of a point M on the line such that the sum $AM + MB$ is minimal.

Problem 2.15 Let ABC be an acute triangle. Find the positions of the points M, N, P on the sides BC, CA, AB, respectively, such that the perimeter of the triangle MNP is minimal.

Problem 2.16 Seven real numbers are given in the interval $(1, 13)$. Prove that at least three of them are the lengths of a triangle's sides.

2.2 An Interesting Locus

Let us consider a triangle ABC and let D be the midpoint of segment BC. It is not difficult to see that

$$[ABD] = [ACD].$$

Moreover, for each point M on the line AD we have

$$[ABM] = [ACM].$$

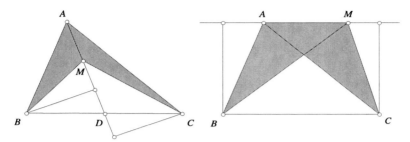

Fig. 2.2

Fig. 2.3

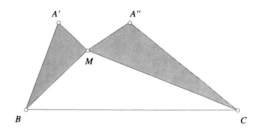

We want to determine the locus of points M such that $[ABM] = [ACM]$. The triangles ABM and ACM have the common side AM; hence their areas are equal if and only if the distances from B and C to AM are equal.

It is not difficult to see that this happens in two situations: AM passes through the midpoint of BC or AM is parallel to BC (see Fig. 2.2). We derive that the locus of points M in the interior of triangle ABC for which $[ABM] = [ACM]$ is the median AD.

Now, let us translate the sides AB and AC to $A'B$ and $A''C$ and ask a similar question. Find the locus of points M such that

$$[A'BM] = [A''CM]$$

(see Fig. 2.3).

Let us rephrase this.

Problem 2.17 Given the quadrilateral $ABCD$ such that AB and CD are not parallel, find the locus of points M inside $ABCD$ for which $[ABM] = [CDM]$.

Solution We apply other translations. Denote by T the point of intersection of the lines AB and CD. We translate the segment AB to TX and the segment CD to TY. It is not difficult to see that $[ABM] = [TXM]$ and $[CDM] = [TYM]$, hence M lies on the median of triangle TXY.

We deduce that the desired locus is a segment: the part of the median of TXY lying inside the quadrilateral $ABCD$ (Fig. 2.4).

Fig. 2.4

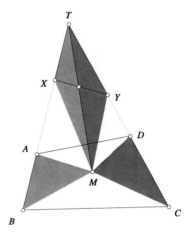

Fig. 2.5

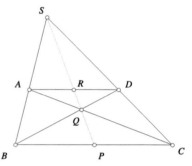

Problem 2.18 Prove that in a trapezoid the midpoints of parallel sides, the point of intersection of the diagonals and the point of intersection of the non-parallel sides are collinear.

Solution This is an immediate application of the above property. We have (with the notations in Fig. 2.5)

$$[ABP] = [CDP], \qquad [ABR] = [CDR],$$

hence R and P lie on the line through the intersection S of AB and CD found in Problem 2.17. All we have to do is to prove that $[ABQ] = [CDQ]$.
But

$$[ABQ] + [BQC] = [ABC] = [DBC] = [CDQ] + [BQC],$$

and we are done.

Problem 2.19 Suppose we are given a positive number k and a quadrilateral $ABCD$ in which AB and CD are not parallel. Find the locus of points M inside $ABCD$ for which $[ABM] + [CDM] = k$.

Fig. 2.6

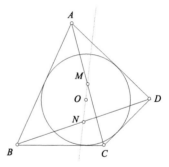

Solution Using the construction from Problem 2.17, we obtain

$$[ABM] + [CDM] = [TXMY] = [TXY] + [XMY].$$

Since X and Y are fixed points, it follows that $[XMY]$ is a constant; therefore the point M lies on a parallel to the line XY. According to the value of k, the locus is either a segment (or a point) or it is the empty set.

Observation If M lies on this parallel, but in the exterior of the quadrilateral, it is not difficult to see that in this case we either have $[ABM] - [CDM] = k$, or $[CDM] - [ABM] = k$.

Problem 2.20 A circle with center O is inscribed in the convex quadrilateral $ABCD$. If M and N are the midpoints of the diagonals AC and BD, prove that points O, M, and N are collinear.

Solution Since a circle is inscribed in quadrilateral $ABCD$ (Fig. 2.6), we know that

$$AB + CD = AD + BC.$$

Multiplying the equality by $r/2$, where r is the radius of the circle, we obtain

$$[OAB] + [OCD] = [OAD] + [OBC].$$

Thus

$$[OAB] + [OCD] = \frac{1}{2}[ABCD].$$

On the other hand,

$$[NAB] + [NCD] = \frac{1}{2}[ABD] + \frac{1}{2}[BCD] = \frac{1}{2}[ABCD].$$

Similarly,

$$[MAB] + [MCD] = \frac{1}{2}[ABCD].$$

It follows that O, M and N belong to the locus of the points X such that

$$[XAB] + [XCD] = \frac{1}{2}[ABCD],$$

and they lie on the same segment if the opposite sides of $ABCD$ are not parallel. If AB is parallel to CD or AD is parallel to BC, the result is trivial.

Let us investigate the following problems:

Problem 2.21 Find the locus of points M in plane of triangle ABC such that $[ABM] = 2[ACM]$.

Problem 2.22 Let D be a point on the side BC of triangle ABC and let M be a point on AD. Prove that

$$\frac{[ABM]}{[ACM]} = \frac{BD}{CD}.$$

Deduce Ceva's theorem: if the segments AD, BE and CF are concurrent then

$$\frac{BD}{CD} \cdot \frac{CE}{AE} \cdot \frac{AF}{BF} = 1.$$

Problem 2.23 Let $ABCD$ be a convex quadrilateral and let M be a point in its interior such that

$$[MAB] = [MBC] = [MCD] = [MDA].$$

Prove that one of the diagonals of $ABCD$ passes through the midpoint of the other diagonal.

Problem 2.24 Let $ABCD$ be a convex quadrilateral. Find the locus of points M in its interior such that

$$[MAB] = 2[MCD].$$

Problem 2.25 Let $ABCD$ be a convex quadrilateral and let $k > 0$ be a real number. Find the locus of points M in its interior such that

$$[MAB] + 2[MCD] = k.$$

Problem 2.26 Let d, d' be two non-parallel lines in the plane and let $k > 0$. Find the locus of points the sum of whose distances to d and d' is equal to k.

Problem 2.27 Let $ABCD$ be a convex quadrilateral and let E and F be the points of intersections of the lines AB, CD and AD, BC, respectively. Prove that the midpoints of the segments AC, BD, and EF are collinear.

Fig. 2.7

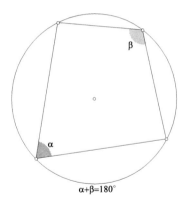

$$\alpha+\beta=180°$$

Fig. 2.8

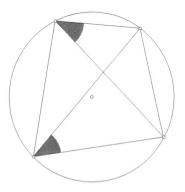

Problem 2.28 In the interior of a quadrilateral $ABCD$, consider a variable point P. Prove that if the sum of distances from P to the sides is constant, then $ABCD$ is a parallelogram.

2.3 Cyclic Quads

A convex quadrilateral is called *cyclic* if its vertices lie on a circle. It is not difficult to see that a necessary and sufficient condition for this is that the sum of the opposite angles of the quadrilateral be equal to 180° (Fig. 2.7).

As a special case, if two opposite angles of the quadrilateral are right angles, then the quadrilateral is cyclic and the diagonal which splits the quadrilateral into two right triangles is a diameter of the circumscribed circle.

Another necessary and sufficient condition is that the angle between one side and a diagonal be equal to the angle between the opposite side and the other diagonal (Fig. 2.8).

Problem 2.29 Let $ABCD$ be a cyclic quadrilateral. Recall that the incenter of a triangle is the intersection of the angles' bisectors. Prove that the incenters of triangles ABC, BCD, CDA and DAB are the vertices of a rectangle.

Fig. 2.9

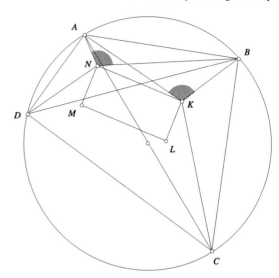

Solution The incenter of a triangle is the intersection of the angle's bisectors. Let K, L, M and N be the four incenters (Fig. 2.9). We then have

$$\angle AKB = 180° - \angle BAC/2 - \angle ABC/2 = 90° + \angle ACB/2.$$

Similarly,

$$\angle ANB = 90° + \angle ADB/2.$$

But since $ABCD$ is cyclic, we have $\angle ACB = \angle ADB$, hence $\angle AKB = \angle ANB$. This means that the quadrilateral $ANKB$ is also cyclic, so

$$\angle NKB = 180° - \angle BAN = 180° - \angle A/2.$$

In the same manner we obtain

$$\angle BKL = 180° - \angle C/2,$$

hence

$$\angle NKL = 360° - (\angle NKB + \angle BKL) = (\angle A + \angle C)/2 = 180°/2 = 90°.$$

Analogously, we prove that the other three angles of $KLMN$ are right angles.

Problem 2.30 In the triangle ABC, the altitude, angle bisector and median from C divide the angle $\angle C$ into four equal angles. Find the angles of the triangle.

Solution No cyclic quads in the text (Fig. 2.10)! However, a solution using some trigonometry is at hand: with usual notations, from the angle bisector's theorem, we have

$$\frac{AE}{EB} = \frac{AC}{CB} = \frac{b}{a},$$

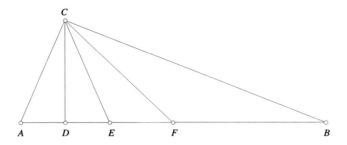

Fig. 2.10

so that

$$\frac{AE}{AB} = \frac{b}{a+b}$$

and

$$AE = \frac{bc}{a+b}.$$

Also,

$$\frac{FB}{FE} = \frac{BC}{CE} = \frac{a}{b}.$$

But

$$FB = \frac{c}{2}, \qquad FE = AF - AE = \frac{c}{2} - \frac{bc}{a+b} = \frac{c(a-b)}{2(a+b)},$$

so that we obtain

$$\frac{\frac{c}{2}}{\frac{c(a-b)}{2(a+b)}} = \frac{a}{b},$$

which is equivalent to

$$\frac{a+b}{a-b} = \frac{a}{b},$$

or $b^2 + 2ab - a^2 = 0$. We get the quadratic equation

$$\left(\frac{b}{a}\right)^2 + 2\frac{b}{a} - 1 = 0,$$

from which we deduce that

$$\frac{b}{a} = \sqrt{2} - 1.$$

On the other hand, using the formula for the length of an angle bisector we have

$$b = CE = \frac{2ab}{a+b}\cos\frac{\angle C}{2}$$

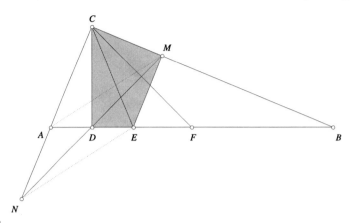

Fig. 2.11

hence

$$\cos\frac{\angle C}{2} = \frac{a+b}{2a} = \frac{1}{2} + \frac{1}{2}\frac{b}{a} = \frac{1}{2} + \frac{1}{2}(\sqrt{2} - 1) = \frac{\sqrt{2}}{2}.$$

We conclude that $\frac{\angle C}{2} = 45°$, so $\angle C = 90°$, $\angle A = 67.5°$ and $\angle B = 22.5°$.

Now, let us try a purely geometric approach. Let D, E and F on AB be the feet of the altitude, angle bisector and median (Fig. 2.11).

Drop a perpendicular from E to BC in the point M and let N be the intersection between MD and AC.

The quadrilateral $CDEM$ is cyclic, since

$$\angle CDE = \angle CME = 90°.$$

Then

$$\angle CMD = \angle CED = \angle CAD.$$

It follows that triangles CMN and CAB are similar.

This implies that CD is the median from C in triangle CMN, hence $AMEN$ is a parallelogram (MN and AE have the same midpoint).

Finally, $AN \parallel ME$ and since $ME \perp BC$, it follows that $AC \perp BC$.

We conclude that $\angle C = 90°$, $\angle A = 67.5°$ and $\angle B = 22.5°$.

Observation Another proof using cyclic quads follows from the lemma below:

Lemma *Let ABC be a triangle inscribed in the circle centered at O such that the angles $\angle B$ and $\angle C$ are acute. If H is its orthocenter, then $\angle BAH = \angle CAO$ (Fig. 2.12).*

Proof Let A' be the point on the circumcircle such that AA' is a diameter. Then $ABA'C$ is cyclic, hence $\angle ABC = \angle AA'C$. It follows that their complementary angles $\angle BAH$ and $\angle CAA'$ are equal as well.   □

Fig. 2.12

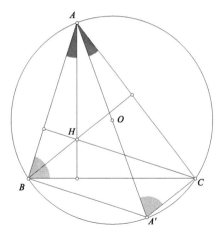

Fig. 2.13

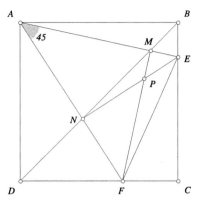

Returning to our problem, note that from the statement it follows that the angles A and B are acute. The orthocenter of the triangle ABC lies on the altitude CD. Because $\angle ACD = \angle BCF$, it follows from the lemma that the circumcenter of the triangle ABC lies on the line CF. On the other hand, the circumcenter lies on the perpendicular bisector of the segment AB. We deduce that the circumcenter is the point F, hence $\angle C = 90°$.

Problem 2.31 Let E and F be two points on the sides BC and CD of the square $ABCD$, such that $\angle EAF = 45°$. Let M and N be the intersections of the diagonal BD with AE and AF, respectively. Let P be the intersection of MF and NE. Prove that AP is perpendicular to EF (Fig. 2.13).

Solution First, observe that $\angle NBE = 45°$, so $ABEN$ is cyclic. This implies that

$$\angle AEN = \angle ABN = 45°,$$

so that triangle ANE is a right isosceles triangle. It follows that EN is perpendicular to AF. Similarly, FM is perpendicular to AE, so P is the orthocenter of triangle AEF. The conclusion is obvious.

Look for the cyclic quads in the following problems.

Problem 2.32 Let D, E, and F be the feet of the altitudes of the triangle ABC. Prove that the altitudes of ABC are the angle bisectors of the triangle DEF.

Problem 2.33 Let $ABCD$ be a convex quadrilateral. Prove that $AB \cdot CD + AD \cdot BC = AC \cdot BD$ if and only if $ABCD$ is cyclic.

Problem 2.34 Let A', B', and C' be points in the interior of the sides BC, CA, and AB of the triangle ABC. Prove that the circumcircles of the triangles $AB'C'$, $BA'C'$, and $CA'B'$ have a common point.

Problem 2.35 Let $ABCD$ be a cyclic quadrilateral. Prove that the orthocenters of the triangles ABC, BCD, CDA and DAB are the vertices of a quadrilateral congruent to $ABCD$ and prove that the centroids of the same triangles are the vertices of a cyclic quadrilateral.

Problem 2.36 Let K, L, M, N be the midpoints of the sides AB, BC, CD, DA, respectively, of a cyclic quadrilateral $ABCD$. Prove that the orthocenters of the triangles AKN, BKL, CLM, DMN are the vertices of a parallelogram.

Problem 2.37 Prove that the perpendiculars dropped from the midpoints of the sides of a convex quadrilateral to the opposite sides are concurrent.

Problem 2.38 In the convex quadrilateral $ABCD$ the diagonals AC and BD intersect at O and are perpendicular. Prove that projections of O on the quadrilateral's sides are the vertices of a cyclic quadrilateral.

2.4 Equiangular Polygons

We call a convex polygon equiangular if its angles are congruent. Thus, an equiangular triangle is an equilateral one, an equiangular quadrilateral is a rectangle (or a square). One interesting property of the equiangular polygons is stated in the following problem.

Problem 2.39 Let P be a variable point in the interior or on the sides of an equiangular polygon. Prove that the sum of distances from P to the polygon's sides is constant.

Fig. 2.14

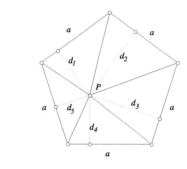

Fig. 2.15

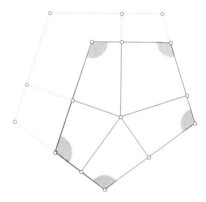

Solution It is not difficult to see that the statement is true for regular polygons. Indeed, connect P to the polygon's vertices and write its area as a sum of areas of the triangles thus obtained (see Fig. 2.14 in the case of a pentagon).

If the polygon's sides have length a and the distances from P to the sides are $d_1, d_2, \ldots, d_n$, respectively, then the area A of the polygon equals $\frac{1}{2} \sum_{k=1}^{n} a d_k$ hence $d_1 + d_2 + \cdots + d_n = \frac{2A}{a}$ is constant.

Now, if the polygon is an arbitrary equiangular one, we can always "expand" it to a regular polygon, adding thus to the sum of distances a constant value (the sum of distances between parallel sides) as seen in Fig. 2.15.

The converse of the above property is true in the case of a triangle. Indeed, if P is a variable point in the interior or on the sides of triangle ABC and the sum of distances from P to triangle's sides is constant, then ABC is an equilateral triangle. We can see this if we move P to A, P to B and P to C and observe that triangle's altitudes must have the same length. It is not difficult to see that this happens if and only if ABC is equilateral.

The converse is not true for a quadrilateral, because a parallelogram has this property and (unless it is a rectangle) it is not an equiangular polygon.

Problem 2.40 Let $a_1, a_2, \ldots, a_n$ be positive real numbers and let ε be the primitive nth root of the unity $\varepsilon = \cos \frac{2\pi}{n} + i \sin \frac{2\pi}{n}$. Prove that if the sides of an equiangular

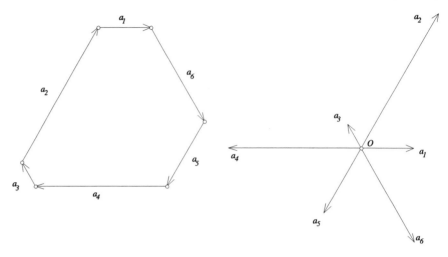

Fig. 2.16

polygon have lengths $a_1, a_2, \ldots, a_n$ (in counter-clockwise order) then

$$a_1 + a_2\varepsilon + a_3\varepsilon^2 + \cdots + a_n\varepsilon^{n-1} = 0.$$

Solution We draw a picture for the case $n = 6$ (the general case is essentially the same).

Consider the polygon's sides as vectors, oriented clockwise (see Fig. 2.16). Then the sum of the vectors equals zero. Now, translate all vectors such that they have the same origin O. If we look at the complex numbers corresponding to their extremities, choosing a_1 on the positive real axis, we see that these are $a_1, a_2\varepsilon, a_3\varepsilon^2, \ldots,$ and $a_n\varepsilon^{n-1}$, respectively. Since the sum of the vectors equals zero, we deduce that $a_1 + a_2\varepsilon + a_3\varepsilon^2 + \cdots + a_n\varepsilon^{n-1} = 0$.

Observation The converse of the statement is generally not true. For instance, if a, b, c, d are the side lengths of a quadrilateral and $a + bi + ci^2 + di^3 = 0$, then $(a - c) + i(b - d) = 0$; this equality is fulfilled if the quadrilateral is a parallelogram (and not necessarily an equiangular quadrilateral, that is, a rectangle). However, from the solution we see that if $a_1, a_2, \ldots, a_n$ are positive numbers and $a_1 + a_2\varepsilon + a_3\varepsilon^2 + \cdots + a_n\varepsilon^{n-1} = 0$, then there exists an equiangular polygon with sides of lengths $a_1, a_2, \ldots, a_n$.

It is worth mentioning that in the case $n = 3$, there exists another necessary and sufficient condition: if a, b, c are the complex numbers corresponding to the (distinct) points A, B, C, then the triangle ABC is equilateral if and only if $a + b\varepsilon + c\varepsilon^2 = 0$, where ε is one of the (non-real) cubic roots of the unity.

Problem 2.41 Prove that if an equiangular hexagon have side lengths $a_1, \ldots, a_6$ (in this order) then $a_1 - a_4 = a_5 - a_2 = a_3 - a_6$.

Fig. 2.17

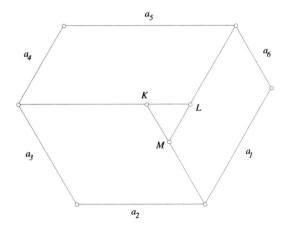

Solution Choose three non-adjacent vertices of the hexagon and draw parallels to the sides through them, as in Fig. 2.17. Suppose that the parallels mutually intersect at the points K, L, M.

Thus, the hexagon is partitioned into three parallelograms and a triangle. Since the hexagon is equiangular, triangle KLM is equilateral. Finally, observe that $LM = a_1 - a_4$, $KL = a_5 - a_2$ and $MK = a_3 - a_6$.

Another solution is possible using the preceding problem. Let $\varepsilon = \cos \frac{2\pi}{6} + i \sin \frac{2\pi}{6}$ be a primitive sixth root of unity. Then

$$a_1 + a_2\varepsilon + a_3\varepsilon^2 + a_4\varepsilon^3 + a_5\varepsilon^4 + a_6\varepsilon^5 = 0.$$

But $\varepsilon^3 = \cos \pi + i \sin \pi = -1$, so $\varepsilon^4 = -\varepsilon$ and $\varepsilon^5 = -\varepsilon^2$. We deduce that

$$(a_1 - a_4) + (a_2 - a_5)\varepsilon + (a_3 - a_6)\varepsilon^2 = 0.$$

On the other hand, since $\varepsilon^3 = -1$ (and $\varepsilon \neq -1$) we see that $\varepsilon^2 - \varepsilon + 1 = 0$. Thus, ε is a common root of the equations $(a_1 - a_4) + (a_2 - a_5)z + (a_3 - a_6)z^2 = 0$ and $z^2 - z + 1 = 0$, both with real coefficients. Since $\varepsilon \notin \mathbf{R}$, it follows that the two equations share another common root $\bar{\varepsilon}$, so the coefficients of the two equations must be proportional; that is, $(a_1 - a_4) = -(a_2 - a_5) = (a_3 - a_6)$, as desired.

Try the following problems.

Problem 2.42 Let $ABCDE$ be an equiangular pentagon whose side lengths are rational numbers. Prove that the pentagon is regular.

Problem 2.43 Prove that p is a prime number if and only if every equiangular polygon with p sides of rational lengths is regular.

Problem 2.44 An equiangular polygon with an odd number of sides is inscribed in a circle. Prove that the polygon is regular.

Fig. 2.18

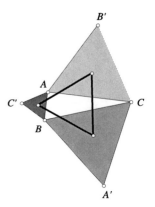

Problem 2.45 Let $a_1, a_2, \ldots, a_n$ be the side lengths of an equiangular polygon. Prove that if $a_1 \geq a_2 \geq \cdots \geq a_n$, then the polygon is regular.

Problem 2.46 The side lengths of an equiangular octagon are rational numbers. Prove that the octagon has a symmetry center.

2.5 More on Equilateral Triangles

One of the most beautiful problems in geometry is the following one.

Problem 2.47 In the exterior of triangle ABC three equilateral triangles ABC', BCA' and CAB' are constructed. Prove that the centroids of these triangles are the vertices of an equilateral triangle.

Solution It is said that the problem was discovered by Napoleon. We do not know the proof that he gave to this statement, but surely it was different from the following one. We use complex numbers. Note that if triangle ABC is equilateral and oriented counter-clockwise and its vertices correspond to the complex numbers a, b, c, then

$$a + b\varepsilon + c\varepsilon^2 = 0, \qquad\qquad (*)$$

where $\varepsilon = \frac{-1+i\sqrt{3}}{2}$ is the cube root of unity which represents a $120°$ counter-clockwise rotation. The converse is also true: suppose $(*)$ holds. Notice that $1 + \varepsilon + \varepsilon^2 = 0$. Thus $\varepsilon^2 = -1 - \varepsilon$ and we obtain $\varepsilon(b - c) = (c - a)$. Hence segment CA is obtained from segment BC by a $120°$ counter-clockwise rotation.

Similarly AB is obtained from CA by a $120°$ counter-clockwise rotation and hence ABC is equilateral and oriented counter-clockwise. Now, let us return to Napoleon's problem.

Assume the triangle ABC is oriented counter-clockwise as in Fig. 2.18. Since $A'CB$, $B'AC$ and $C'BA$ are oriented counter-clockwise and are equilateral, we have

$$a' + c\varepsilon + b\varepsilon^2 = 0, \qquad b' + a\varepsilon + c\varepsilon^2 = 0, \qquad c' + b\varepsilon + a\varepsilon^2 = 0,$$

respectively. The complex numbers corresponding to the centroids of triangles $A'BC$, $AB'C$ and ABC' are

$$a'' = \frac{1}{3}(a' + b + c),$$

$$b'' = \frac{1}{3}(a + b' + c)$$

and

$$c'' = \frac{1}{3}(a + b + c'),$$

respectively.

We have to check that $a'' + b''\varepsilon + c''\varepsilon^2 = 0$. Observe that

$$a'' = \frac{1}{3}(-c\varepsilon - b\varepsilon^2 + b + c) = \frac{1}{3}\left(b(1 - \varepsilon^2) + c(1 - \varepsilon)\right),$$

$$b''\varepsilon = \frac{1}{3}(a - a\varepsilon - c\varepsilon^2 + c)\varepsilon = \frac{1}{3}\left(a(\varepsilon - \varepsilon^2) + c(\varepsilon - 1)\right),$$

$$c''\varepsilon^2 = \frac{1}{3}(a + b - b\varepsilon - a\varepsilon^2)\varepsilon^2 = \frac{1}{3}\left(a(\varepsilon^2 - \varepsilon) + b(\varepsilon^2 - 1)\right).$$

Adding up the three equalities gives the desired result.

The reader who wants a "proof without words" for Napoleon's problem should carefully examine Fig. 2.19 (which may also be of some interest to carpet manufacturers...).

Problem 2.48 In the exterior of the acute triangle ABC, three equilateral triangles ABC', BCA' and CAB' are constructed. Prove that the segments AA', BB' and CC' are concurrent. Also, prove that the circumcircles of the equilateral triangles pass through the same point (Fig. 2.20).

Solution We start with another problem: find the point T in the interior of triangle ABC for which the sum $TA + TB + TC$ is minimal. Rotate clockwise the points A and T around C with 60°, to B' and T', respectively (see Fig. 2.21). Then $B'T' = AT$ and since triangle CTT' is equilateral, $TT' = TC$.

We deduce that the sum $TA + TB + TC$ is equal to $BT + TT' + T'B'$, and the latter is minimal when the points B, T, T' and B' are collinear. Thus, the point T for which the sum $TA + TB + TC$ is minimal lies on BB'. Similarly, we can prove that

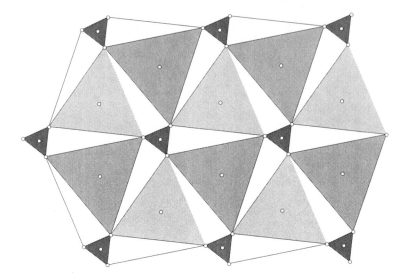

Fig. 2.19

Fig. 2.20

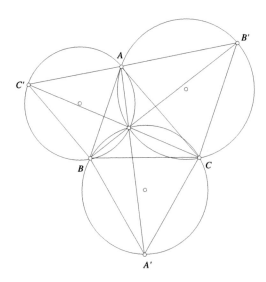

T lies on AA' and CC', so the three segments are concurrent. Also (see Fig. 2.22) notice that we have

$$\angle ATC = \angle CT'B' = 180° - \angle CT'T = 120°$$

and

$$\angle BTC = 180° - \angle CTT' = 120°.$$

Fig. 2.21

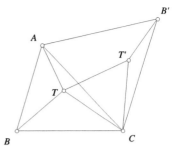

Fig. 2.22

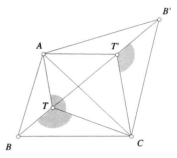

We deduce that the point T is the point in the interior of the triangle with the property that

$$\angle ATB = \angle BTC = \angle CTA = 120°$$

(this point is called Toricelli's point). In this case the quadrilaterals $ATCB'$, $ATBC'$ and $BCTA'$ are cyclic, so the circumcircles of the equilateral triangles pass through T.

Here are some more problems.

Problem 2.49 Let M be a point in the interior of the equilateral triangle ABC and let A', B', C' be its projections onto the sides BC, CA and AB, respectively.

Prove that the sum of lengths of the inradii of triangles MAC', MBA' and MCB' equals the sum of lengths of the inradii of triangles MAB', MBC' and MCA' (Fig. 2.23).

Problem 2.50 Let I be the incenter of triangle ABC. It is known that for every point $M \in (AB)$, one can find the points $N \in (BC)$ and $P \in (AC)$ such that I is the centroid of triangle MNP. Prove that ABC is an equilateral triangle.

Problem 2.51 Let ABC be an acute triangle. The interior bisectors of the angles $\angle B$ and $\angle C$ meet the opposite sides in the points L and M, respectively. Prove that there exists a point K in the interior of the side BC such that triangle KLM is equilateral if and only if $\angle A = 60°$.

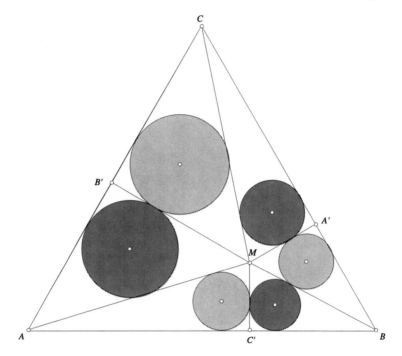

Fig. 2.23

Problem 2.52 Let $P_1 P_2 \ldots P_n$ be a convex polygon with the following property: for any two vertices P_i and P_j, there exists a vertex P_k such that $\angle P_i P_k P_j = 60°$. Prove that the polygon is an equilateral triangle.

Problem 2.53 From a point on the circumcircle of an equilateral triangle ABC parallels to the sides BC, CA and AB are drawn, intersecting the sides CA, AB and BC at the points M, N, P, respectively. Prove that the points M, N, P are collinear.

Problem 2.54 Let P be a point on the circumcircle of an equilateral triangle ABC. Prove that the projections of any point Q on the lines PA, PB and PC are the vertices of an equilateral triangle.

2.6 The "Carpets" Theorem

We start with a textbook problem.

Problem 2.55 Let M and N be the midpoints of the sides AB and BC of the square $ABCD$. Let $P = AN \cap DM$, $Q = AN \cap CM$ and $R = CM \cap DN$.
    Prove the equality

$$[AMP] + [BMQN] + [CNR] = [DPQR].$$

Fig. 2.24

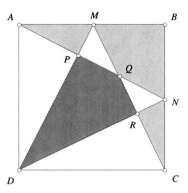

Solution Let us try brute force (Fig. 2.24). Assume with no loss of generality that the square's area equals 1. Since the figure is symmetric with respect to BD, it suffices to prove that

$$[AMP] + [BMQ] = [DPQ].$$

The triangles AQD and NQB are similar, so

$$\frac{AQ}{QN} = \frac{AD}{BN} = 2.$$

It follows that

$$[AQB] = \frac{2}{3}[ABN] = \frac{1}{6}$$

and then

$$[BMQ] = \frac{1}{12}.$$

Let $N' = AN \cap CD$. Then triangles AMP and $N'DP$ are also similar and

$$\frac{PM}{PD} = \frac{AM}{DN'} = \frac{1}{4}.$$

Consequently,

$$[AMP] = \frac{1}{5}[AMD] = \frac{1}{20}.$$

Finally, it is easy to see that

$$\frac{PQ}{AD} = \frac{4}{15},$$

and

$$[PQD] = \frac{4}{15}[AND] = \frac{2}{15}.$$

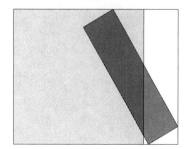

Fig. 2.25

Fig. 2.26

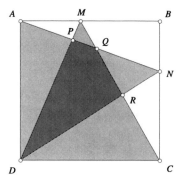

Now,

$$\frac{1}{12} + \frac{1}{20} = \frac{2}{15}$$

and we are done.

However, this is an inelegant solution. Fortunately, there exists a much simpler one, with no computations and that works for arbitrary points M and N on the respective sides!

First, let us see the "carpets" theorem. Suppose that the floor of a rectangular room is completely covered by a collection of nonoverlapping carpets. If we move one of the carpets, then clearly the overlapping area is equal to the uncovered area of the floor (see Fig. 2.25).

Of course, the shape of the room or the shape of the carpets are irrelevant.

Now, let us return to Problem 2.55, taking M and N at arbitrary positions:

The "room" is $ABCD$ and the "carpets" are triangles ADN and CDM (Fig. 2.26). Since

$$[ADN] = [CDM] = \frac{1}{2}[ABCD],$$

the two carpets would completely cover the room if they did not overlap. Hence the area of the overlapping surface, that is, $[DPQR]$, equals the area of the uncovered

Fig. 2.27

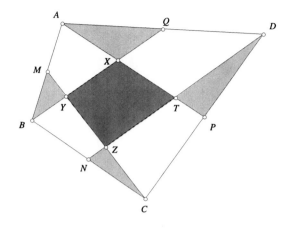

Fig. 2.28

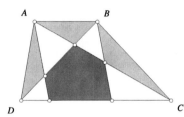

surface, i.e.

$$[AMP] + [BMQN] + [CNR].$$

Arrange some carpets to solve the following problems.

Problem 2.56 Let $ABCD$ be a parallelogram. The points M, N and P are chosen on the segments BD, BC and CD, respectively, such that $CNMP$ is a parallelogram. Let $E = AN \cap BD$ and $F = AP \cap BD$. Prove that

$$[AEF] = [DFP] + [BEN].$$

Problem 2.57 Consider the quadrilateral $ABCD$. The points M, N, P and Q are the midpoints of the sides AB, BC, CD and DA (Fig. 2.27).

Let $X = AP \cap BQ$, $Y = BQ \cap CM$, $Z = CM \cap DN$ and $T = DN \cap AP$. Prove that

$$[XYZT] = [AQX] + [BMY] + [CNZ] + [DPT].$$

Problem 2.58 Through the vertices of the smaller base AB of the trapezoid $ABCD$ two parallel lines are drawn, intersecting the segment CD. These lines and the trapezoid's diagonals divide it into seven triangles and a pentagon (see Fig. 2.28).

Show that the area of the pentagon equals the sum of areas of the three triangles sharing a common side with the trapezoid.

Fig. 2.29

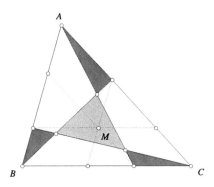

Problem 2.59 Let M be a point in the interior of triangle ABC. Three lines are drawn through M, parallel to triangle's sides, determining three trapezoids. One draws a diagonal in each trapezoid such that they have no common endpoints, dividing thus ABC into seven parts, four of them being triangles (see Fig. 2.29).

Prove that the area of one of the four triangles equals the sum of the areas of the other three.

2.7 Quadrilaterals with an Inscribed Circle

Everybody knows that in any triangle one can inscribe a circle whose center is at the intersection point of the angles' bisectors.

What can we say about a quadrilateral? If there exists a circle touching all the quadrilateral's sides, then its center is equidistant from them, hence it lies on all four angle bisectors. We deduce that a necessary and sufficient condition for the existence of an inscribed circle is that the quadrilateral's angle bisectors are concurrent (in fact, this works for arbitrary convex polygons).

This does not happen in every quadrilateral. We can always draw a circle tangent to three of the four sides (its center being the point of intersection of two of the bisectors) (Fig. 2.30).

If three of the four angle bisectors meet at one point, it is easy to see that the fourth one will also pass through that point and that a circle can be inscribed in the quadrilateral.

Another necessary and sufficient condition for the existence of an inscribed circle in a quadrilateral is given by the following theorem, due to Pithot.

Theorem *Let $ABCD$ be a convex quadrilateral. There exists a circle inscribed in $ABCD$ if and only if:*

$$AB + CD = AD + BC.$$

Proof Suppose there exists a circle inscribed in the quadrilateral, touching the sides AB, BC, CD and DA at the points K, L, M, N, respectively. Then, since the tan-

Fig. 2.30

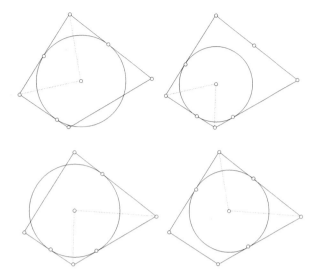

Fig. 2.31

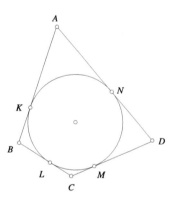

gents from a point to a circle have equal lengths, we have $AK = AN$, $BK = BL$, $CM = CL$ and $DM = DN$ (Fig. 2.31).

If we add up these equalities, we obtain the desired result.   □

Conversely, suppose $AB + CD = AD + BC$. Draw a circle tangent to AB, BC and CD. If the circle is not tangent to AD, draw from A the tangent to the circle and let E be the point of intersection with CD. Suppose, for instance, that E lies in the interior of CD (Fig. 2.32).

Since the circle is inscribed in the quadrilateral $ABCE$, we have $AB + CE = AE + BC$. On the other hand, from the hypothesis, we have $AB + CD = AD + BC$, or $AB + CE + ED = AD + BC$. From these, we derive $ED + AE = AD$, which is impossible. It follows that the circle is also tangent to AD, hence it is inscribed in $ABCD$.

If E lies outside the line segment CD, the proof is almost identical.

Fig. 2.32

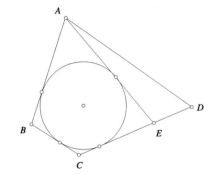

Fig. 2.33

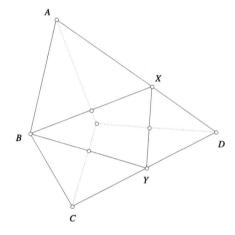

Another nice proof of the converse is the following: if $AB = AD$ then $BC = CD$ and the conclusion is immediate. Suppose, with no loss of generality, that $AB < AD$, and let $X \in AD$ be such that $AB = AX$. Let $Y \in CD$ be such that $DX = DY$. Since $AB + CD = AD + BC$, it follows that $CY = BC$ (Fig. 2.33).

Thus, the triangles ABX, DXY and CYB are isosceles, so the perpendicular bisectors of the sides BX, XY and YB of triangle BXY are also the angle bisectors of $\angle A, \angle D$, and $\angle C$ of the quadrilateral. Since the perpendicular bisectors of the sides of a triangle are concurrent, we obtain the desired conclusion.

Problem 2.60 Prove that if in the quadrilateral $ABCD$ is inscribed a circle with center O, then the sum of the angles $\angle AOB$ and $\angle COD$ equals $180°$.

Problem 2.61 Let $ABCD$ be a quadrilateral with an inscribed circle. Prove that the circles inscribed in triangles ABC and ADC are tangent to each other.

Problem 2.62 Let $ABCD$ be a convex quadrilateral. Suppose that the lines AB and CD intersect at E and the lines AD and BC intersect at F, such that the points E and F lie on opposite sides of the line AC. Prove that the following statements are equivalent:

Fig. 2.34

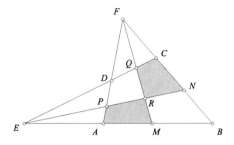

   (i) a circle is inscribed in $ABCD$;

  (ii) $BE + BF = DE + DF$;

 (iii) $AE - AF = CE - CF$.

Problem 2.63 Let $ABCD$ be a convex quadrilateral. Suppose that the lines AB and CD intersect at E and the lines AD and BC intersect at F. Let M and N be two arbitrary points on the line segments AB and BC, respectively. The line EN intersects AF and MF at P and R. The line MF intersects CE at Q. Prove that if the quadrilaterals $AMRP$ and $CNRQ$ have inscribed circles, then $ABCD$ has an inscribed circle (Fig. 2.34).

Problem 2.64 The points A_1, A_2, C_1 and C_2 are chosen in the interior of the sides CD, BC, AB and AD of the convex quadrilateral $ABCD$. Denote by M the point of intersection of the lines AA_2 and CC_1 and by N the point of intersection of the lines AA_1 and CC_2. Prove that if one can inscribe circles in three of the four quadrilaterals $ABCD, A_2BC_1M, AMCN$ and A_1NC_2D, then a circle can be also inscribed in the fourth one.

Problem 2.65 A line cuts a quadrilateral with an inscribed circle into two polygons with equal areas and equal perimeters. Prove that the line passes through the center of the inscribed circle.

Problem 2.66 In the convex quadrilateral $ABCD$ we have $\angle B = \angle C = 120°$, and

$$AB^2 + BC^2 + CD^2 = AD^2.$$

Prove that $ABCD$ has an inscribed circle.

Problem 2.67 Let $ABCD$ be a quadrilateral circumscribed about a circle, whose interior and exterior angles are at least $60°$. Prove that

$$\frac{1}{3}\left|AB^3 - AD^3\right| \le \left|BC^3 - CD^3\right| \le 3\left|AB^3 - AD^3\right|.$$

When does equality hold?

2.8 Dr. Trig Learns Complex Numbers

It is known that every complex number $z = a + bi$ can be written in the form

$$z = r(\cos\theta + i\sin\theta),$$

where $r = \sqrt{a^2 + b^2}$ is the absolute value of z and θ is its argument. For instance, $i = \cos\frac{\pi}{2} + i\sin\frac{\pi}{2}$, $1 + i = \frac{\sqrt{2}}{2}(\cos\frac{\pi}{4} + i\sin\frac{\pi}{4})$, etc. Also, if $z = r(\cos\theta + i\sin\theta)$, then a simple inductive argument proves that

$$z^n = r^n(\cos n\theta + i\sin n\theta)$$

(known as de Moivre's formula).

Now, if $z = \cos\theta + i\sin\theta$ (that is, its absolute value equals 1) then

$$\frac{1}{z} = \frac{1}{\cos\theta + i\sin\theta} = \frac{\cos\theta - i\sin\theta}{\cos^2\theta + \sin^2\theta} = \cos\theta - i\sin\theta.$$

We obtain the useful formulas

$$\cos\theta = \frac{1}{2}\left(z + \frac{1}{z}\right), \qquad \sin\theta = \frac{1}{2i}\left(z - \frac{1}{z}\right).$$

Moreover, using de Moivre's formula, we have

$$\cos n\theta = \frac{1}{2}\left(z^n + \frac{1}{z^n}\right), \qquad \sin n\theta = \frac{1}{2i}\left(z^n - \frac{1}{z^n}\right).$$

These formulas may be very useful in solving a lot of Dr. Trig's problems. We might even forget some of his formulas. For instance, we have

$$\cos 2\theta = \frac{1}{2}\left(z^2 + \frac{1}{z^2}\right) = \frac{1}{2}\left(z + \frac{1}{z}\right)^2 - 1 = 2\left[\frac{1}{2}\left(z + \frac{1}{z}\right)\right]^2 - 1 = 2\cos^2\theta - 1.$$

Also

$$\left[\frac{1}{2}\left(z + \frac{1}{z}\right)\right]^3 = \frac{1}{8}\left(z^3 + 3z + \frac{3}{z} + \frac{1}{z^3}\right) = \frac{1}{8}\left(z^3 + \frac{1}{z^3}\right) + \frac{3}{8}\left(z + \frac{1}{z}\right).$$

We deduce

$$\cos^3\theta = \frac{1}{4}\cos 3\theta + \frac{3}{4}\cos\theta,$$

a formula sometimes written as

$$\cos 3\theta = 4\cos^3\theta - 3\cos\theta.$$

Now, let us put some of this at work.

Problem 2.68 Let a, b, c be real numbers such that

$$\cos a + \cos b + \cos c = \sin a + \sin b + \sin c = 0.$$

Prove that

$$\cos 2a + \cos 2b + \cos 2c = \sin 2a + \sin 2b + \sin 2c = 0.$$

Solution Let $x = \cos a + i \sin a$, $y = \cos b + i \sin b$ and $z = \cos c + i \sin c$. From the hypothesis we have

$$x + y + z = 0,$$

and also

$$\frac{1}{x} + \frac{1}{y} + \frac{1}{z} = (\cos a - i \sin a) + (\cos b - i \sin b) + (\cos c - i \sin c) = 0.$$

It follows that $xy + xz + yz = 0$ so

$$x^2 + y^2 + z^2 = (x + y + z)^2 - 2(xy + xz + yz) = 0.$$

But then

$$\cos 2a + \cos 2b + \cos 2c + i(\sin 2a + \sin 2b + \sin 2c) = 0,$$

and we are done.

Problem 2.69 Prove the equality

$$\cos \frac{\pi}{7} + \cos \frac{3\pi}{7} + \cos \frac{5\pi}{7} = \frac{1}{2}.$$

Solution Let

$$z = \cos \frac{\pi}{7} + i \sin \frac{\pi}{7}.$$

Then $z^7 = \cos \pi + i \sin \pi = -1$, hence $z^7 + 1 = 0$. On the other hand, we have

$$\cos \frac{\pi}{7} + \cos \frac{3\pi}{7} + \cos \frac{5\pi}{7} = \frac{1}{2}\left(z + \frac{1}{z}\right) + \frac{1}{2}\left(z^3 + \frac{1}{z^3}\right) + \frac{1}{2}\left(z^5 + \frac{1}{z^5}\right)$$

$$= \frac{z^{10} + z^8 + z^6 + z^4 + z^2 + 1}{2z^5}.$$

Since $z^7 + 1 = 0$, we have $z^{10} = -z^3$ and $z^8 = -z$. It follows that

$$z^{10} + z^8 + z^6 + z^4 + z^2 + 1$$
$$= z^6 + z^4 - z^3 + z^2 - z + 1$$
$$= z^6 - z^5 + z^4 - z^3 + z^2 - z + 1 + z^5 = \frac{z^7 + 1}{z + 1} + z^5 = z^5,$$

hence

$$\cos\frac{\pi}{7}+\cos\frac{3\pi}{7}+\cos\frac{5\pi}{7}=\frac{z^5}{2z^5}=\frac{1}{2}.$$

Problem 2.70 Find a closed form for the sum

$$S_n=\sin a+\sin 2a+\cdots+\sin na.$$

Solution Let us ask for more: we will find a closed form for both S_n and

$$C_n=\cos a+\cos 2a+\cdots+\cos na.$$

Let $z=\cos a+i\sin a$. Then

$$C_n+iS_n=z+z^2+\cdots+z^n=z\frac{z^n-1}{z-1}.$$

Since $\cos x-1=-2\sin^2\frac{x}{2}$ and $\sin x=2\sin\frac{x}{2}\cos\frac{x}{2}$, we obtain

$$\frac{z^n-1}{z-1}=\frac{\cos na+i\sin na-1}{\cos a+i\sin a-1}=\frac{-2\sin^2\frac{na}{2}+2i\sin\frac{na}{2}\cos\frac{na}{2}}{-2\sin^2\frac{a}{2}+2i\sin\frac{a}{2}\cos\frac{a}{2}}$$

$$=\frac{\sin\frac{na}{2}}{\sin\frac{a}{2}}\left(\frac{\cos\frac{na}{2}+i\sin\frac{na}{2}}{\cos\frac{a}{2}+i\sin\frac{a}{2}}\right)=\frac{\sin\frac{na}{2}}{\sin\frac{a}{2}}\left(\cos\frac{(n-1)a}{2}+i\sin\frac{(n-1)a}{2}\right).$$

Thus

$$C_n+iS_n=(\cos a+i\sin a)\frac{\sin\frac{na}{2}}{\sin\frac{a}{2}}\left(\cos\frac{(n-1)a}{2}+i\sin\frac{(n-1)a}{2}\right)$$

$$=\frac{\sin\frac{na}{2}}{\sin\frac{a}{2}}\left(\cos\frac{(n+1)a}{2}+i\sin\frac{(n+1)a}{2}\right).$$

Finally

$$S_n=\frac{\sin\frac{na}{2}\sin\frac{(n+1)a}{2}}{\sin\frac{a}{2}},\qquad C_n=\frac{\sin\frac{na}{2}\cos\frac{(n+1)a}{2}}{\sin\frac{a}{2}}.$$

Now, help Dr. Trig to solve the following problems.

Problem 2.71 Let a,b,c be real numbers such that

$$\cos a+\cos b+\cos c=\sin a+\sin b+\sin c=0.$$

Prove that

$$\cos(a+b+c)=\frac{1}{3}(\cos 3a+\cos 3b+\cos 3c),$$

$$\sin(a+b+c) = \frac{1}{3}(\sin 3a + \sin 3b + \sin 3c).$$

Problem 2.72 Find the value of the product $\cos 20° \cos 40° \cos 80°$.

Problem 2.73 Prove that

$$\frac{1}{\cos 6°} + \frac{1}{\sin 24°} + \frac{1}{\sin 48°} = \frac{1}{\sin 12°}.$$

Problem 2.74 Prove that

$$\cos \frac{2\pi}{7} + \cos \frac{4\pi}{7} + \cos \frac{6\pi}{7} + \frac{1}{2} = 0.$$

Problem 2.75 Prove the equality

$$\sin \frac{\pi}{n} \sin \frac{2\pi}{n} \cdots \sin \frac{(n-1)\pi}{n} = \frac{\sqrt{n}}{2^{n-1}}.$$

Problem 2.76 Solve the equation

$$\sin x + \sin 2x + \sin 3x = \cos x + \cos 2x + \cos 3x.$$

Problem 2.77 Prove that

$$\cos \frac{\pi}{5} = \frac{1+\sqrt{5}}{4}.$$

Chapter 3
Number Theory and Combinatorics

3.1 Arrays of Numbers

Many Olympiad problems refer to arrays of numbers. Let us start with some examples.

Problem 3.1 The numbers $1, 2, \ldots, n^2$ are arranged in an $n \times n$ array in the following way:

1	2	3	...	n
$n+1$	$n+2$	$n+3$	...	$2n$
$\vdots$				$\vdots$
$n^2 - n + 1$	$n^2 - n + 2$	$n^2 - n + 3$	...	n^2

Pick n numbers from the array such that any two numbers are in different rows and different columns. Find the sum of these numbers.

Solution If we denote by a_{ij} the number in the ith row and jth column then

$$a_{ij} = (i-1)n + j$$

for all $i, j = 1, 2, \ldots, n$. Because any two numbers are in different rows and different columns, it follows that from each row and each column exactly one number is chosen. Let $a_{1j_1}, a_{2j_2}, \ldots, a_{nj_n}$ be the chosen numbers, where $j_1, j_2, \ldots, j_n$ is a permutation of indices $1, 2, \ldots, n$. We have

$$\sum_{k=1}^{n} a_{kj_k} = \sum_{k=1}^{n} \left((k-1)n + j_k\right) = n \sum_{k=1}^{n} (k-1) + \sum_{k=1}^{n} j_k.$$

But

$$\sum_{k=1}^{n} j_k = \frac{n(n+1)}{2}$$

T. Andreescu, B. Enescu, *Mathematical Olympiad Treasures*,
DOI 10.1007/978-0-8176-8253-8_3, © Springer Science+Business Media, LLC 2011

since $j_1, j_2, \ldots, j_n$ is a permutation of indices $1, 2, \ldots, n$. It follows that

$$\sum_{k=1}^{n} a_{k j_k} = \frac{n(n^2 + 1)}{2}.$$

Problem 3.2 The entries of an $n \times n$ array of numbers are denoted by x_{ij}, with $1 \le i, j \le n$. For all i, j, k, $1 \le i, j, k \le n$ the following equality holds

$$x_{ij} + x_{jk} + x_{ki} = 0.$$

Prove that there exist numbers $t_1, t_2, \ldots, t_n$ such that

$$x_{ij} = t_i - t_j$$

for all i, j, $1 \le i, j \le n$.

Solution Setting $i = j = k$ in the given condition yields $3x_{ii} = 0$, hence $x_{ii} = 0$ for all i, $1 \le i \le n$. For $k = j$ we obtain

$$x_{ij} + x_{jj} + x_{ji} = 0,$$

hence $x_{ij} = -x_{ji}$, for all i, j. Now fix i and j and add up the equalities

$$x_{ij} + x_{jk} + x_{ki} = 0$$

for $k = 1, 2, \ldots, n$. It follows that

$$n x_{ij} + \sum_{k=1}^{n} x_{jk} + \sum_{k=1}^{n} x_{ki} = 0$$

or

$$n x_{ij} + \sum_{k=1}^{n} x_{jk} - \sum_{k=1}^{n} x_{ik} = 0.$$

If we define

$$t_i = \frac{1}{n} \sum_{k=1}^{n} x_{ik},$$

we obtain

$$x_{ij} = t_i - t_j,$$

for all i, j, as desired.

Problem 3.3 Prove that among any 10 entries of the table

$$
\begin{array}{cccccc}
0 & 1 & 2 & 3 & \ldots & 9 \\
9 & 0 & 1 & 2 & \ldots & 8 \\
8 & 9 & 0 & 1 & \ldots & 7 \\
\vdots & & & & & \\
1 & 2 & 3 & 4 & \ldots & 0
\end{array}
$$

situated in different rows and different columns, at least two are equal.

Solution Denote by a_{ij} the entries of the table, $1 \le i, j \le 10$ and observe that $a_{ij} = a_{hk}$ if and only if $i - j \equiv h - k \pmod{10}$. Let a_{ij_i}, $i = 1, 2, \ldots, 10$ be 10 entries situated in different rows and different columns. If these entries are all different, then the differences $i - j_i$ give distinct residues mod 10, hence

$$
\sum_{i=1}^{10}(i - j_i) \equiv 0 + 1 + \cdots + 9 \equiv 5 \pmod{10}.
$$

On the other hand, since $j_1, j_2, \ldots, j_{10}$ is a permutation of the indices $1, 2, \ldots, 10$, we have

$$
\sum_{i=1}^{10}(i - j_i) = \sum_{i=1}^{10} i - \sum_{i=1}^{10} j_i = 0,
$$

hence $0 \equiv 5 \pmod{10}$, a contradiction.

Here are some proposed problems.

Problem 3.4 Prove that the sum of any n entries of the table

$$
\begin{array}{ccccc}
1 & \frac{1}{2} & \frac{1}{3} & \cdots & \frac{1}{n} \\
\frac{1}{2} & \frac{1}{3} & \frac{1}{4} & \cdots & \frac{1}{n+1} \\
\vdots & & & & \\
\frac{1}{n} & \frac{1}{n+1} & \frac{1}{n+2} & \cdots & \frac{1}{2n-1}
\end{array}
$$

situated in different rows and different columns is not less than 1.

Problem 3.5 The entries of an $n \times n$ array of numbers are denoted by a_{ij}, $1 \le i, j \le n$. The sum of any n entries situated on different rows and different columns is the same. Prove that there exist numbers $x_1, x_2, \ldots, x_n$ and $y_1, y_2, \ldots, y_n$, such that

$$
a_{ij} = x_i + y_j,
$$

for all i, j.

Problem 3.6 In an $n \times n$ array of numbers all rows are different (two rows are different if they differ in at least one entry). Prove that there is a column which can be deleted in such a way that the remaining rows are still different.

Problem 3.7 The positive integers from 1 to n^2 ($n \geq 2$) are placed arbitrarily on squares of an $n \times n$ chessboard. Prove that there exist two adjacent squares (having a common vertex or a common side) such that the difference of the numbers placed on them is not less than $n + 1$.

Problem 3.8 A positive integer is written in each square of an $n^2 \times n^2$ chess board. The difference between the numbers in any two adjacent squares (sharing an edge) is less than or equal to n. Prove that at least $\lfloor \frac{n}{2} \rfloor + 1$ squares contain the same number.

Problem 3.9 The numbers $1, 2, \ldots, 100$ are arranged in the squares of an 10×10 table in the following way: the numbers $1, \ldots, 10$ are in the bottom row in increasing order, numbers $11, \ldots, 20$ are in the next row in increasing order, and so on. One can choose any number and two of its neighbors in two opposite directions (horizontal, vertical, or diagonal). Then either the number is increased by 2 and its neighbors are decreased by 1, or the number is decreased by 2 and its neighbors are increased by 1. After several such operations the table again contains all the numbers $1, 2, \ldots, 100$. Prove that they are in the original order.

Problem 3.10 Prove that one cannot arrange the numbers from 1 to 81 in a 9×9 table such that for each i, $1 \leq i \leq 9$ the product of the numbers in row i equals the product of the numbers in column i.

Problem 3.11 The entries of a matrix are integers. Adding an integer to all entries on a row or on a column is called an operation. It is given that for infinitely many integers N one can obtain, after a finite number of operations, a table with all entries divisible by N. Prove that one can obtain, after a finite number of operations, the zero matrix.

3.2 Functions Defined on Sets of Points

Several Olympiad problems deal with functions defined on certain sets of points. These problems are interesting in that they combine both geometrical and algebraic ideas.

Problem 3.12 Let $n > 2$ be an integer and $f : P \to \mathbf{R}$ a function defined on the set of points in the plane, with the property that for any regular n-gon $A_1 A_2 \ldots A_n$,

$$f(A_1) + f(A_2) + \cdots + f(A_n) = 0.$$

Prove that f is the zero function.

Solution Let A be an arbitrary point. Consider a regular n-gon $AA_1A_2\ldots A_{n-1}$. Let k be an integer, $0 \le k \le n-1$. A rotation with center A of angle $\frac{2k\pi}{n}$ sends the polygon $AA_1A_2\ldots A_{n-1}$ to $A_{k0}A_{k1}\ldots A_{k,n-1}$, where $A_{k0} = A$ and A_{ki} is the image of A_i, for all $i = 1, 2, \ldots, n-1$.

From the condition of the statement, we have

$$\sum_{k=0}^{n-1}\sum_{i=0}^{n-1} f(A_{ki}) = 0.$$

Observe that in the sum the number $f(A)$ appears n times, therefore

$$nf(A) + \sum_{k=0}^{n-1}\sum_{i=1}^{n-1} f(A_{ki}) = 0.$$

On the other hand, we have

$$\sum_{k=0}^{n-1}\sum_{i=1}^{n-1} f(A_{ki}) = \sum_{i=1}^{n-1}\sum_{k=0}^{n-1} f(A_{ki}) = 0,$$

since the polygons $A_{0i}A_{1i}\ldots A_{n-1,i}$ are all regular n-gons. From the two equalities above we deduce $f(A) = 0$, hence f is the zero function.

Problem 3.13 Let $n \ge 4$ be an integer and let $p \ge 2n - 3$ be a prime number. Let S be a set of n points in the plane, no three of which are collinear, and let $f : S \to \{0, 1, \ldots, p-1\}$ be a function such that

1. there exists a unique point $A \in S$ such that $f(A) = 0$;
2. if $C(X, Y, Z)$ is the circle determined by the distinct points $X, Y, Z \in S$, then

$$\sum_{P \in S \cap C(X,Y,Z)} f(P) \equiv 0 \ (\mathrm{mod}\ p).$$

Prove that all points of S lie on a circle.

Solution Suppose there exist points B and C such that no other point of S lies on $C(A, B, C)$. We have

$$f(B) + f(C) \equiv 0 \ (\mathrm{mod}\ p),$$

so, if $f(B) = i \neq 0$, then $f(C) = p - i$. Let

$$\sigma = \sum_{X \in S} f(X).$$

If a number of b circles pass through A, B and other points of S and a number of c circles pass through A, C and other points of S, applying the condition from the

Fig. 3.1

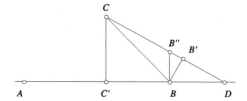

hypothesis to all these circles, we obtain

$$\sigma + (b-1)i \equiv 0 \pmod{p},$$
$$\sigma + (c-1)(p-i) \equiv 0 \pmod{p},$$

hence

$$b + -2 \equiv 0 \pmod{p}.$$

Since $1 \le b, c \le n-2$, we have $2 \le b+c \le 2n-4 < p$, hence $b = c = 1$, which is a contradiction.

It follows that for any points $B, C \in S$, there exists at least one more point of S lying on $C(A, B, C)$. This implies the fact that all the points of S lie on a circle. Indeed, consider an inversion I of pole A. The set of points $S - \{A\}$ is transformed into the set of points $I(S - \{A\}) = N$ and the circles $C(A, X, Y)$, with $X, Y \in S$ are transformed into lines $I_X I_Y$ through points of N. The above condition then reduces to the following: for any two points I_X, I_Y of N, there exists at least one other point of N lying on the line $I_X I_Y$. We can prove that all points of N are collinear, whence the points of S lie on a circle.

Indeed, suppose the points are not collinear and choose points A, B, C such that the distance from C to the line AB is minimal. Let C' be the projection of C on the line AB. From the hypothesis it follows that there exists another point $D \in N$ on the line AB. The point C' divides the line AB into two half lines and at least two of the points A, B, D lie in one of these half lines. Suppose these points are B and D located as in Fig. 3.1.

If B'' is a point on CD such that BB'' is parallel to CC' and B' is the projection of B on CD, it is not difficult to see that $BB' < BB'' < CC'$, contradicting the minimality of CC'.

Here are more examples.

Problem 3.14 Let D be the union of $n \ge 1$ concentric circles in the plane. Suppose that the function $f : D \to D$ satisfies

$$d\big(f(A), f(B)\big) \ge d(A, B)$$

for every $A, B \in D$ ($d(M, N)$ is the distance between the points M and N).
    Prove that

$$d\big(f(A), f(B)\big) = d(A, B)$$

for every $A, B \in D$.

Problem 3.15 Let S be a set of $n \geq 4$ points in the plane, such that no three of them are collinear and not all of them lie on a circle. Find all functions $f : S \rightarrow \mathbf{R}$ with the property that for any circle C containing at least three points of S,

$$\sum_{P \in C \cap S} f(P) = 0.$$

Problem 3.16 Let P be the set of all points in the plane and L be the set of all lines of the plane. Find, with proof, whether there exists a bijective function $f : P \rightarrow L$ such that for any three collinear points A, B, C, the lines $f(A)$, $f(B)$ and $f(C)$ are either parallel or concurrent.

Problem 3.17 Let S be the set of interior points of a sphere and C be the set of interior points of a circle. Find, with proof, whether there exists a function $f : S \rightarrow C$ such that $d(A, B) \leq d(f(A), f(B))$, for any points $A, B \in S$.

Problem 3.18 Let S be the set of all polygons in the plane. Prove that there exists a function $f : S \rightarrow (0, +\infty)$ such that

1. $f(P) < 1$, for any $P \in S$;
2. If $P_1, P_2 \in S$ have disjoint interiors and $P_1 \cup P_2 \in S$, then $f(P_1 \cup P_2) = f(P_1) + f(P_2)$.

3.3 Count Twice!

Many interesting results can be obtained by counting the elements of a set in two different ways. Moreover, counting twice can be a good problem solving strategy. Let us consider the following examples.

Problem 3.19 In how many ways can a committee of k persons with a chairman be chosen from a set of n people?

Solution We can choose the k members of the committee in $\binom{n}{k}$ ways and then the chairman in k ways, so that the answer is $k\binom{n}{k}$. On the other hand, we can first choose the chairman (this can be done in n ways) and next the rest of $k-1$ members from the remaining $n-1$ persons, leading to a total of $n\binom{n-1}{k-1}$ possibilities. Thus, we obtain the following identity:

$$k\binom{n}{k} = n\binom{n-1}{k-1}.$$

Problem 3.20 Consider a finite sequence of real numbers the sum of any 7 consecutive terms is negative and the sum of any 11 consecutive terms is positive. Find the greatest number of terms of such a sequence.

Solution It is not difficult to see that such a sequence cannot have 77 or more terms. For that, we add up in two ways the first 77 terms, obtaining a contradiction:

$$(x_1 + x_2 + \cdots + x_7) + (x_8 + x_9 + \cdots + x_{14}) + \cdots + (x_{71} + x_{72} + \cdots + x_{77}) < 0,$$

$$(x_1 + x_2 + \cdots + x_{11}) + (x_{12} + x_{13} + \cdots + x_{22}) + \cdots$$

$$+ (x_{67} + x_{68} + \cdots + x_{77}) > 0.$$

In fact, the number of terms is much less. The sequence cannot contain more than 16 terms. To prove that, we refine our double counting technique.

Suppose the sequence has at least 17 terms. We can arrange them in a 7×11 table as follows:

$$\begin{pmatrix} x_1 & x_2 & \ldots & x_{10} & x_{11} \\ x_2 & x_3 & \ldots & x_{11} & x_{12} \\ \vdots & \vdots & & \vdots & \vdots \\ x_6 & x_7 & \ldots & x_{15} & x_{16} \\ x_7 & x_8 & \ldots & x_{16} & x_{17} \end{pmatrix}$$

Now, add up all the entries in two ways: by rows and by columns. We again reach a contradiction.

Finally, there is an example of such a sequence with 16 terms:

$$5, 5, -13, 5, 5, 5, -13, 5, 5, -13, 5, 5, 5, -13, 5, 5.$$

Problem 3.21 Prove the equality

$$\sum_{d|n} \varphi(d) = n,$$

where φ denotes Euler's totient function.

Solution The Euler's totient function $\varphi(n)$ denotes the number of positive integers less than or equal to n and relatively prime to n, that is, the number of elements of the set

$$\{k \,|\, 1 \le k \le n, \text{ and } \gcd(k, n) = 1\}.$$

If the prime decomposition of the positive integer n is

$$n = p_1^{a_1} p_2^{a_2} \cdots p_k^{a_k},$$

then

$$\varphi(n) = n\left(1 - \frac{1}{p_1}\right)\left(1 - \frac{1}{p_2}\right) \cdots \left(1 - \frac{1}{p_k}\right),$$

and, using this formula, a computational proof of our assertion is possible.

However, we will present a simpler proof, based on counting arguments.

Consider the fractions

$$\frac{1}{n}, \frac{2}{n}, \frac{3}{n}, \ldots, \frac{n-1}{n}, \frac{n}{n}.$$

Obviously, there are n such fractions (this is not a silly observation; just the first way to count them).

Now, consider the same fractions in their lowest terms. Obviously, their denominators are divisors of n. How many of them have still the denominator equal to n? Clearly, those whose numerator was relatively prime to n, that is, $\varphi(n)$. How many have now the denominator equal to some d, where d is a divisor of n? In every such fraction, the numerator has to be less than or equal to d and relatively prime to d, hence there are $\varphi(d)$ such fractions. We thus obtained

$$\sum_{d \mid n} \varphi(d) = n,$$

as desired.

Try counting twice in the following problems.

Problem 3.22 Find how many committees with a chairman can be chosen from a set of n persons. Derive the identity

$$\binom{n}{1} + 2\binom{n}{2} + 3\binom{n}{3} + \cdots + n\binom{n}{n} = n2^{n-1}.$$

Problem 3.23 In how many ways can one choose k balls from a set containing $n-1$ red balls and a blue one? Derive the identity

$$\binom{n}{k} = \binom{n-1}{k} + \binom{n-1}{k-1}.$$

Problem 3.24 Let S be a set of n persons such that:

(i) any person is acquainted to exactly k other persons in S;
(ii) any two persons that are acquainted have exactly l common acquaintances in S;
(iii) any two persons that are not acquainted have exactly m common acquaintances in S.

Prove that

$$m(n - k) - k(k - l) + k - m = 0.$$

Problem 3.25 Let n be an odd integer greater than 1 and let $c_1, c_2, \ldots, c_n$ be integers. For each permutation $a = (a_1, a_2, \ldots, a_n)$ of $\{1, 2, \ldots, n\}$, define

$$S(a) = \sum_{i=1}^{n} c_i a_i.$$

Prove that there exist permutations $a \neq b$ of $\{1, 2, \ldots, n\}$ such that $n!$ is a divisor of $S(a) - S(b)$.

Problem 3.26 Let $a_1 \leq a_2 \leq \cdots \leq a_n = m$ be positive integers. Denote by b_k the number of those a_i for which $a_i \geq k$. Prove that

$$a_1 + a_2 + \cdots + a_n = b_1 + b_2 + \cdots + b_m.$$

Problem 3.27 In how many ways can one fill a $m \times n$ table with ± 1 such that the product of the entries in each row and each column equals -1?

Problem 3.28 Let n be a positive integer. Prove that

$$\sum_{k=0}^{n} \binom{n}{k} \binom{n+k}{k} = \sum_{k=0}^{n} 2^k \binom{n}{k}^2.$$

Problem 3.29 Prove that

$$1^2 + 2^2 + \cdots + n^2 = \binom{n+1}{2} + 2\binom{n+1}{3}.$$

Problem 3.30 Let n and k be positive integers and let S be a set of n points in the plane such that

(a) no three points of S are collinear, and
(b) for every point P of S there are at least k points of S equidistant from P.

Prove that

$$k < \frac{1}{2} + \sqrt{2 \cdot n}.$$

Problem 3.31 Prove that

$$\tau(1) + \tau(2) + \cdots + \tau(n) = \left\lfloor \frac{n}{1} \right\rfloor + \left\lfloor \frac{n}{2} \right\rfloor + \cdots + \left\lfloor \frac{n}{n} \right\rfloor,$$

where $\tau(k)$ denotes the number of divisors of the positive integer k.

Problem 3.32 Prove that

$$\sigma(1) + \sigma(2) + \cdots + \sigma(n) = \left\lfloor \frac{n}{1} \right\rfloor + 2\left\lfloor \frac{n}{2} \right\rfloor + \cdots + n\left\lfloor \frac{n}{n} \right\rfloor,$$

where $\sigma(k)$ denotes the sum of divisors of the positive integer k.

Problem 3.33 Prove that

$$\varphi(1)\left\lfloor \frac{n}{1} \right\rfloor + \varphi(2)\left\lfloor \frac{n}{2} \right\rfloor + \cdots + \varphi(n)\left\lfloor \frac{n}{n} \right\rfloor = \frac{n(n+1)}{2},$$

where φ denotes Euler's totient function.

3.4 Sequences of Integers

Sequences of integers are a favorite of Olympiad problem writers, since such sequences involve several different mathematical concepts, including, for example, algebraic techniques, recursive relations, divisibility and primality.

Problem 3.34 Consider the sequence $(a_n)_{n\geq 1}$ defined by $a_1 = a_2 = 1$, $a_3 = 199$ and

$$a_{n+1} = \frac{1989 + a_n a_{n-1}}{a_{n-2}}$$

for all $n \geq 3$. Prove that all the terms of the sequence are positive integers.

Solution We have

$$a_{n+1} a_{n-2} = 1989 + a_n a_{n-1}.$$

Replacing n by $n - 1$ yields

$$a_n a_{n-3} = 1989 + a_{n-1} a_{n-2}$$

and we obtain

$$a_{n+1} a_{n-2} - a_n a_{n-1} = a_n a_{n-3} - a_{n-1} a_{n-2}.$$

This is equivalent to

$$a_{n-2}(a_{n+1} + a_{n-1}) = a_n(a_{n-1} + a_{n-3})$$

or

$$\frac{a_{n+1} + a_{n-1}}{a_n} = \frac{a_{n-1} + a_{n-3}}{a_{n-2}}$$

for all $n \geq 4$. If n is even, we obtain

$$\frac{a_{n+1} + a_{n-1}}{a_n} = \frac{a_{n-1} + a_{n-3}}{a_{n-2}} = \cdots = \frac{a_3 + a_1}{a_2} = 200,$$

while if n is odd,

$$\frac{a_{n+1} + a_{n-1}}{a_n} = \frac{a_{n-1} + a_{n-3}}{a_{n-2}} = \cdots = \frac{a_4 + a_2}{a_3} = 11.$$

It follows that

$$a_{n+1} = \begin{cases} 200a_n - a_{n-1}, & \text{if } n \text{ is even;} \\ 11a_n - a_{n-1}, & \text{if } n \text{ is odd.} \end{cases}$$

An inductive argument shows that all a_n are positive integers.

Problem 3.35 Let a, b, c be positive real numbers. The sequence $(a_n)_{n\geq 1}$ is defined by $a_1 = a, a_2 = b$ and

$$a_{n+1} = \frac{a_n^2 + c}{a_{n-1}}$$

for all $n \geq 2$. Prove that the terms of the sequence are all positive integers if and only if a, b and $\frac{a^2+b^2+c}{ab}$ are positive integers.

Solution Clearly, all the terms of the sequence are positive numbers. Write the recursive relation as

$$a_{n+1}a_{n-1} = a_n^2 + c.$$

Replacing n by $n - 1$ yields

$$a_n a_{n-2} = a_{n-1}^2 + c,$$

and by subtracting the two equalities we deduce

$$a_{n-1}(a_{n+1} + a_{n-1}) = a_n(a_n + a_{n-2}).$$

Therefore

$$\frac{a_{n+1} + a_{n-1}}{a_n} = \frac{a_n + a_{n-2}}{a_{n-1}}$$

for all $n \geq 3$. It follows that the sequence $b_n = \frac{a_{n+1}+a_{n-1}}{a_n}$ is constant, say $b_n = k$, for all $n \geq 2$. Then the sequence (a_n) satisfies the recursive relation

$$a_{n+1} = ka_n - a_{n-1}$$

for all $n \geq 2$ and since $a_3 = \frac{b^2+c}{a} = kb - a$, we derive that

$$k = \frac{a^2 + b^2 + c}{ab}.$$

Now, if a, b and k are positive integers, it follows inductively that a_n is a positive integer for all $n \geq 1$. Conversely, suppose that a_n is a positive integer for all $n \geq 1$. Then a, b are positive integers and $k = \frac{a_3+a}{b}$ is a rational number. Let $k = \frac{p}{q}$, where p and q are relatively prime positive integers. We want to prove that $q = 1$. Suppose that $q > 1$. From the recursive relation we obtain

$$q(a_{n+1} + a_{n-1}) = pa_n,$$

and hence q divides a_n for all $n \geq 2$. We prove by induction on s that q^s divides a_n for all $n \geq s + 1$. We have seen that this is true for $s = 1$. Suppose q^{s-1} divides a_n for all $n \geq s$. We have

$$a_{n+2} = \frac{p}{q}a_{n+1} - a_n,$$

which is equivalent to

$$\frac{a_{n+2}}{q^{s-1}} = p\frac{a_{n+1}}{q^s} - \frac{a_n}{q^{s-1}}.$$

If $n \geq s$, then q^{s-1} divides a_n and a_{n+2}, hence q^s divides a_{n+1}. It follows that q^s divides a_n for all $n \geq s+1$. Finally, we have

$$a_{s+2} = \frac{a_{s+1}^2 + c}{a_s}$$

or

$$c = a_s a_{s+2} - a_s^2,$$

which implies that c is divisible by $q^{2(s-1)}$ for all $s \geq 1$. Because $c > 0$, this is a contradiction.

Problem 3.36 Consider the sequence $(a_n)_{n\geq 0}$ given by the following relation: $a_0 = 4$, $a_1 = 22$, and for all $n \geq 2$,

$$a_n = 6a_{n-1} - a_{n-2}.$$

Prove that there exist sequences of positive integers $(x_n)_{n\geq 0}$, $(y_n)_{n\geq 0}$ such that

$$a_n = \frac{y_n^2 + 7}{x_n - y_n},$$

for all $n \geq 0$.

Solution Observe that a_n is an even positive integer for all n and that the sequence $(a_n)_{n\geq 0}$ is increasing. The last assertion follows inductively if we write the recursive relation under the form

$$a_n - a_{n-1} = 5(a_{n-1} - a_{n-2}) + 4a_{n-2}.$$

Define

$$x_n = \frac{a_n + a_{n-1}}{2}, \qquad y_n = \frac{a_n - a_{n-1}}{2}$$

with $x_0 = 3$, $y_0 = 1$. Observe that

$$x_n = 3x_{n-1} + 4y_{n-1}$$

and

$$y_n = 2x_{n-1} + 3y_{n-1}.$$

We have $a_n = x_n + y_n$, hence it is sufficient to prove that

$$x_n + y_n = \frac{y_n^2 + 7}{x_n - y_n}.$$

or

$$x_n^2 = 2y_n^2 + 7$$

for all n. We use induction. The equality is true for $n = 0$. Suppose that $x_{n-1}^2 = 2y_{n-1}^2 + 7$. Then

$$x_n^2 - 2y_n^2 - 7 = (3x_{n-1} + 4y_{n-1})^2 - 2(2x_{n-1} + 3y_{n-1})^2 - 7$$
$$= x_{n-1}^2 - 2y_{n-1}^2 - 7 = 0,$$

which proves our claim.

Try your hand at the following problems.

Problem 3.37 Prove that there exist sequences of odd positive integers $(x_n)_{n\geq 3}$, $(y_n)_{n\geq 3}$ such that

$$7x_n^2 + y_n^2 = 2^n$$

for all $n \geq 3$.

Problem 3.38 Let $x_1 = x_2 = 1$, $x_3 = 4$ and

$$x_{n+3} = 2x_{n+2} + 2x_{n+1} - x_n$$

for all $n \geq 1$. Prove that x_n is a square for all $n \geq 1$.

Problem 3.39 The sequence $(a_n)_{n\geq 0}$ is defined by $a_0 = a_1 = 1$ and

$$a_{n+1} = 14a_n - a_{n-1}$$

for all $n \geq 1$. Prove that the number $2a_n - 1$ is a square for all $n \geq 0$.

Problem 3.40 The sequence $(x_n)_{n\geq 1}$ is defined by $x_1 = 0$ and

$$x_{n+1} = 5x_n + \sqrt{24x_n^2 + 1}$$

for all $n \geq 1$. Prove that all x_n are positive integers.

Problem 3.41 Let $(a_n)_{n\geq 1}$ be an increasing sequence of positive integers such that

1. $a_{2n} = a_n + n$ for all $n \geq 1$;
2. if a_n is a prime, then n is a prime.

Prove that $a_n = n$, for all $n \geq 1$.

Problem 3.42 Let $a_0 = a_1 = 1$ and $a_{n+1} = 2a_n - a_{n-1} + 2$, for all $n \geq 1$. Prove that

$$a_{n^2+1} = a_{n+1}a_n,$$

for all $n \geq 0$.

Problem 3.43 Let $a_0 = 1$ and $a_{n+1} = a_0 \cdots a_n + 4$, for all $n \geq 0$. Prove that

$$a_n - \sqrt{a_{n+1}} = 2,$$

for all $n \geq 1$.

Problem 3.44 The sequence $(x_n)_{n\geq 1}$ is defined by $x_1 = 1$, $x_2 = 3$ and $x_{n+2} = 6x_{n+1} - x_n$, for all $n \geq 1$. Prove that $x_n + (-1)^n$ is a perfect square, for all $n \geq 1$.

Problem 3.45 Let $(a_n)_{n\geq 1}$ be a sequence of non-negative integers such that $a_n \geq a_{2n} + a_{2n+1}$, for all $n \geq 1$. Prove that for any positive integer N we can find N consecutive terms of the sequence, all equal to zero.

3.5 Equations with Infinitely Many Solutions

In this section we have selected several Diophantine equations having an infinite number of solutions.

Problem 3.46 Find integer solutions for the equation

$$4x^2 + 9y^2 = 72z^2.$$

Solution We first notice that x must be divisible by 3 and that y is an even integer. Setting $x = 3u$ and $y = 2v$ yields

$$36u^2 + 36v^2 = 72z^2$$

or

$$2z^2 = u^2 + v^2.$$

We deduce that u, v have the same parity, so that $u + v$ and $u - v$ are even integers. It follows that

$$z^2 = \left(\frac{u+v}{2}\right)^2 + \left(\frac{u-v}{2}\right)^2.$$

This is the well-known Pythagorean equation, whose solutions are

$$z = k\left(m^2 + n^2\right), \qquad \frac{u+v}{2} = 2kmn, \qquad \frac{u-v}{2} = k\left(m^2 - n^2\right),$$

yielding

$$x = 3k\left(2mn + m^2 - n^2\right), \qquad y = 2k\left(2mn - m^2 + n^2\right), \qquad z = k\left(m^2 + n^2\right).$$

Problem 3.47 Prove that the equation

$$x^2 + (x + 1)^2 = y^2$$

has infinitely many solutions in positive integers.

Solution Observe that $x = 3$, $y = 5$ is a solution. We define the sequences $(x_n)_{n\geq1}$, $(y_n)_{n\geq1}$ by

$$x_{n+1} = 3x_n + 2y_n + 1,$$
$$y_{n+1} = 4x_n + 3y_n + 2,$$

with $x_1 = 3$ and $y_1 = 5$. Suppose that (x_n, y_n) is a solution of the equation. We claim that (x_{n+1}, y_{n+1}) is also a solution and since the sequences are clearly increasing, the equation has infinitely many solutions.

Indeed, we have

$$x_{n+1}^2 + (x_{n+1} + 1)^2 = (3x_n + 2y_n + 1)^2 + (3x_n + 2y_n + 2)^2.$$

Using the equality

$$x_n^2 + (x_n + 1)^2 = y_n^2$$

yields

$$(3x_n + 2y_n + 1)^2 + (3x_n + 2y_n + 2)^2 = (4x_n + 3y_n + 2)^2$$

and hence

$$x_{n+1}^2 + (x_{n+1} + 1)^2 = y_{n+1}^2,$$

thus proving the claim.

Here are some proposed problems.

Problem 3.48 Find all triples of integers (x, y, z) such that

$$x^2 + xy = y^2 + xz.$$

Problem 3.49 Let n be an integer number. Prove that the equation

$$x^2 + y^2 = n + z^2$$

has infinitely many integer solutions.

Problem 3.50 Let m be a positive integer. Find all pairs of integers (x, y) such that

$$x^2(x^2 + y) = y^{m+1}.$$

Problem 3.51 Let m be a positive integer. Find all pairs of integers (x, y) such that

$$x^2(x^2 + y^2) = y^{m+1}.$$

Problem 3.52 Find all non-negative integers a, b, c, d, n such that

$$a^2 + b^2 + c^2 + d^2 = 7 \cdot 4^n.$$

Problem 3.53 Show that there are infinitely many systems of positive integers (x, y, z, t) which have no common divisor greater than 1 and such that

$$x^3 + y^3 + z^2 = t^4.$$

Problem 3.54 Let $k \geq 6$ be an integer number. Prove that the system of equations

$$\begin{cases} x_1 + x_2 + \cdots + x_{k-1} = x_k, \\ x_1^3 + x_2^3 + \cdots + x_{k-1}^3 = x_k, \end{cases}$$

has infinitely many integral solutions.

Problem 3.55 Solve in integers the equation

$$x^2 + y^2 = (x - y)^3.$$

Problem 3.56 Let a and b be positive integers. Prove that if the equation

$$ax^2 - by^2 = 1$$

has a solution in positive integers, then it has infinitely many solutions.

Problem 3.57 Prove that the equation

$$\frac{x+1}{y} + \frac{y+1}{x} = 4$$

has infinitely many solutions in positive integers.

Problem 3.58 Prove that the equation

$$x^3 + y^3 - 2z^3 = 6(x + y + 2z)$$

has infinitely many solutions in positive integers.

3.6 Equations with No Solutions

One of the most famous problems in number theory is Fermat's Last Theorem. The theorem states that for $n \geq 3$, the equation

$$x^n + y^n = z^n$$

has no non-zero integer solutions. Hundreds of articles and books were written about this theorem, which was finally proved by Andrew Wiles in 1994.

For our purposes, however, we will consider much simpler Diophantine equations with no solutions.

Problem 3.59 Prove that the equation

$$x^3 + y^3 + z^3 = 2002$$

has no solutions in integers.

Solution We first notice that a cube of an integer is congruent to either $0, 1$ or $-1 \pmod 9$. Then the sum of two cubes can be congruent to $0, 1, -1, 2$ or -2 and the sum of three cubes to $0, 1, -1, 2, -2, 3$ or -3. Since 2002 is congruent to $4 \pmod 9$ the equation has no solutions.

Problem 3.60 Prove that the equation

$$4x^3 - 7y^3 = 2003$$

has no solutions in integers.

Solution Because $2003 = 1 + 7 \cdot 286$, the equation is equivalent to

$$4x^3 - 1 = 7(y^3 + 286).$$

Notice that if x is an integer, then $x^3 \equiv 0, 1$ or $6 \pmod 7$, so that $4x^3 \equiv 0, 4$ or $3 \pmod 7$. We conclude that $4x^3 - 1$ is not divisible by 7, hence the equation has no integer solutions.

Problem 3.61 Prove that the equation

$$x^3 + y^4 = 7$$

has no solution in integers.

Solution The residues of a cube modulo 13 are $0, 1, 5, 8$ or 12 while the residues of a fourth power are $0, 1, 3$ or 9 modulo 13. This leads us to conclude that the equation $x^3 + y^4 \equiv 7 \pmod{13}$ cannot hold.

Here are some proposed problems.

Problem 3.62 Prove that the equation

$$4xy - x - y = z^2$$

has no positive integer solutions.

Problem 3.63 Prove that the equation

$$6(6a^2 + 3b^2 + c^2) = 5d^2$$

has no solution in non-zero integers.

Problem 3.64 Prove that the system of equations

$$\begin{cases} x^2 + 6y^2 = z^2, \\ 6x^2 + y^2 = t^2 \end{cases}$$

has no positive integer solutions.

Problem 3.65 Let k and n be positive integers, with $n > 2$. Prove that the equation

$$x^n - y^n = 2^k$$

has no positive integer solutions.

Problem 3.66 Prove that the equation

$$\frac{x^{2000} - 1}{x - 1} = y^2$$

has no positive integer solutions.

Problem 3.67 Prove that the equation

$$4(x_1^4 + x_2^4 + \cdots + x_{14}^4) = 7(x_1^3 + x_2^3 + \cdots + x_{14}^3)$$

has no solution in positive integers.

Problem 3.68 Prove that the equation

$$x^2 + y^2 + z^2 = 2011^{2011} + 2012$$

has no solution in integers.

Problem 3.69 Prove that the system

$$x^6 + x^3 + x^3 y + y = 147^{157},$$
$$x^3 + x^3 y + y^2 + y + z^9 = 157^{147}$$

has no solution in integers x, y, and z.

Problem 3.70 Prove that the equation

$$x^5 + y^5 + 1 = (x + 2)^5 + (y - 3)^5$$

has no solution in integers.

Problem 3.71 Prove that the equation

$$x^5 = y^2 + 4$$

has no solution in integers.

Problem 3.72 Prove that the equation

$$x^3 - 3xy^2 + y^3 = 2891$$

has no solution in integers.

3.7 Powers of 2

Many problems deal in one way or another with powers of two and related ideas like binary representation and divisibility.

Problem 3.73 Prove that the numbers $F_n = 2^{2^n} + 1$, $n = 0, 1, 2, \ldots$, are pairwise coprime.

Solution Let $m > n$ and $d = \gcd(F_m, F_n)$. Now, $F_m - 2 = 2^{2^m} - 1 = (2^{2^{n+1}})^{2^{m-n-1}} - 1$ is divisible by $2^{2^{n+1}} - 1$, since $x^k - 1$ is always divisible by $x - 1$. But $2^{2^{n+1}} - 1 = (2^{2^n} - 1)(2^{2^n} + 1) = (2^{2^n} - 1) F_n$, hence F_n divides $F_m - 2$. We deduce that d is a divisor of 2 and, since all F_n are odd numbers, d must be equal to 1, thus proving the claim. Observe that the above property incidentally also proves that there exist infinitely many prime numbers.

Problem 3.74 The sequence $(a_n)_{n \geq 1}$ satisfies $a_1 = 2$, $a_2 = 3$ and, for each $n \geq 2$, either $a_{n+1} = 2a_{n-1}$ or $a_{n+1} = 3a_n - 2a_{n-1}$. Prove that no integer between 1600 and 2000 can be a term of the sequence.

Solution Clearly, the sequence is not uniquely determined by the above conditions. A short analysis shows that the first 6 terms could be

(a) $2, 3, 4, 6, 8, 12$;
(b) $2, 3, 4, 6, 10, 12$;
(c) $2, 3, 4, 6, 10, 18$;
(d) $2, 3, 5, 6, 10, 12$;
(e) $2, 3, 5, 6, 10, 18$;
(f) $2, 3, 5, 6, 8, 12$;
(g) $2, 3, 5, 9, 10, 18$;
(h) $2, 3, 5, 9, 10, 12$;
(i) $2, 3, 5, 9, 17, 18$;
(j) $2, 3, 5, 9, 17, 33$;

The key observation is that each of the numbers $2, 3, 4, 5, 6, 8, 9, 10, 12, 17, 18$ and 33 can be written as a sum of two powers of 2 (or, equivalently, that their binary representation contains exactly two 1's). Thus, we may conjecture that $a_n = 2^{x_n} + 2^{y_n}$, for some integers x_n, y_n and try induction. Unfortunately, the attempt fails unless we observe something more. Let us take, for instance, the case (h) and write the terms as sums of powers of 2: $2 = 2^0 + 2^0$, $3 = 2^0 + 2^1$, $5 = 2^0 + 2^2$, $9 = 2^0 + 2^3$, $10 = 2^1 + 2^3$, $12 = 2^2 + 2^3$. If you do not see the pattern, let us take case (c): $2 = 2^0 + 2^0$, $3 = 2^0 + 2^1$, $4 = 2^1 + 2^1$, $6 = 2^1 + 2^2$, $10 = 2^1 + 2^3$, $18 = 2^1 + 2^4$.

We can see that at each step, exactly one of the numbers x_n and y_n increases with one unit. Now, we can state a stronger induction hypothesis which is easier to prove (this is not uncommon!): for each n, $a_n = 2^{x_n} + 2^{y_n}$, with $x_n \leq x_{n+1}$, $y_n \leq y_{n+1}$ and $x_n + y_n + 1 = x_{n+1} + y_{n+1}$. This obviously holds for $n = 1, 2$. Suppose it holds for all $k \leq n$ and look at a_{n+1}.

If $a_{n+1} = 2a_{n-1}$, then

$$a_{n+1} = 2\left(2^{x_{n-1}} + 2^{y_{n-1}}\right) = 2^{x_{n-1}+1} + 2^{y_{n-1}+1}.$$

If $x_n = x_{n-1} + 1$, $y_n = y_{n-1}$, then let $x_{n+1} = x_n$ and $y_{n+1} = y_n + 1$. We obtain $a_{n+1} = 2^{x_{n+1}} + 2^{y_{n+1}}$. If $x_n = x_{n-1}$, $y_n = y_{n-1} + 1$, then let $x_{n+1} = x_n + 1$ and $y_{n+1} = y_n$. Also, we get $a_{n+1} = 2^{x_{n+1}} + 2^{y_{n+1}}$.

If $a_{n+1} = 3a_n - 2a_{n-1}$, then

$$a_{n+1} = 3\left(2^{x_n} + 2^{y_n}\right) - 2\left(2^{x_{n-1}} + 2^{y_{n-1}}\right).$$

If $x_n = x_{n-1} + 1$, $y_n = y_{n-1}$, then

$$a_{n+1} = 3\left(2^{x_{n-1}+1} + 2^{y_{n-1}}\right) - 2\left(2^{x_{n-1}} + 2^{y_{n-1}}\right) = 2^{x_{n-1}+2} + 2^{y_{n-1}},$$

so that if we take $x_{n+1} = x_n + 1$ and $y_{n+1} = y_n$, we obtain $a_{n+1} = 2^{x_{n+1}} + 2^{y_{n+1}}$. The case $x_n = x_{n-1}$, $y_n = y_{n-1} + 1$ is similar thus our assertion is proved.

Finally, we notice that the binary representations of the numbers 1600 and 2000 are 11001000000 and 11111010000, respectively. Clearly, no number between them has a binary representation with only two 1's. In fact, we can replace 1600 and 2000 with 1537 and 2047.

Here are some proposed problems:

Problem 3.75 Let n be a positive integer such that $2^n + 1$ is a prime number. Prove that $n = 2^k$, for some integer k.

Problem 3.76 Let n be a positive integer such that $2^n - 1$ is a prime number. Prove that n is a prime number.

Problem 3.77 Prove that the number $2^{1992} - 1$ can be written as a product of 6 integers greater than 2^{248}.

Problem 3.78 Determine the remainder of $3^{2^n} - 1$ when divided by 2^{n+3}.

Problem 3.79 Prove that for each n, there exists a number A_n, divisible by 2^n, whose decimal representation contains n digits, each of them equal to 1 or 2.

Problem 3.80 Using only the digits 1 and 2, suppose we write down numbers with 2^n digits such that the digits of every two of them differ in at least 2^{n-1} places. Prove that no more than 2^{n+1} such numbers exist.

Problem 3.81 Does there exist a natural number N which is a power of 2 whose digits (in the decimal representation) can be permuted to form a different power of 2?

Problem 3.82 For a positive integer N, let $s(N)$ the sum of its digits, in the decimal representation. Prove that there are infinitely many n for which $s(2^n) > s(2^{n+1})$.

Problem 3.83 Find all integers of the form 2^n (where n is a natural number) with the property that deleting the first digit of its decimal representation again yields a power of 2.

Problem 3.84 Let $a_0 = 0$, $a_1 = 1$ and, for $n \geq 2$, $a_n = 2a_{n-1} + a_{n-2}$. Prove that a_n is divisible by 2^k if and only if n is divisible by 2^k.

Problem 3.85 If $A = \{a_1, a_2, \ldots, a_p\}$ is a set of real numbers such that $a_1 > a_2 > \cdots > a_p$, we define

$$s(A) = \sum_{k=1}^{p} (-1)^{k-1} a_k.$$

Let M be a set of n positive integers. Prove that $\sum_{A \subseteq M} s(A)$ is divisible by 2^{n-1}.

Problem 3.86 Find all positive integers a, b, such that the product

$$(a + b^2)(b + a^2)$$

is a power of 2.

Problem 3.87 Let $f(x) = 4^x + 6^x + 9^x$. Prove that if m and n are positive integers, then $f(2^m)$ divides $f(2^n)$ whenever $m \leq n$.

Problem 3.88 Show that, for any fixed integer $n \geq 1$, the sequence

$$2, 2^2, 2^{2^2}, 2^{2^{2^2}}, \ldots (\text{mod } n)$$

is eventually constant.

3.8 Progressions

We call a sequence $(a_n)_{n \geq 1}$ of real numbers an arithmetical progression (sequence) if there exists a real number r such that

$$a_{n+1} = a_n + r,$$

for every $n \geq 1$. Then r is called the common difference of the sequence and the terms of the progression are uniquely determined by a_1 and r. It is easy to see that

$$a_n = a_1 + (n - 1)r,$$

for every $n \geq 1$ and that

$$S_n = a_1 + a_2 + \cdots + a_n = \frac{a_1 + a_n}{2} \cdot n.$$

Also, notice that for every $n \geq 2$,

$$a_n = \frac{a_{n-1} + a_{n+1}}{2}.$$

A sequence $(b_n)_{n \geq 1}$ of non-zero real numbers is called a geometrical sequence (progression) if there exists $q \neq 0$ such that for every $n \geq 1$,

$$b_{n+1} = b_n \cdot q.$$

Here, q is called the common ratio of the progression.
We have $b_n = b_1 \cdot q^{n-1}$ and

$$b_1 + b_2 + \cdots + b_n = b_1 \frac{q^n - 1}{q - 1},$$

for every $n \geq 1$, if $q \neq 1$.
We will focus on integer progressions.

Problem 3.89 Prove that if an arithmetical progression of positive integers contains a square, then it contains infinitely many squares.

Solution Assume with no loss of generality that the first term is a square, say a^2, and let r be the common difference. Hence the progression is

$$a^2, a^2 + r, a^2 + 2r, \ldots .$$

Eventually we get to

$$a^2 + (2a + r)r = (a + r)^2.$$

By repeating the argument we can find infinitely many squares among the progression's terms.

Problem 3.90 Prove that one can eliminate some terms of an arithmetical progression of positive integers in such a way that the remaining terms form a geometric progression.

Solution Let a be the first term and r the common difference of the arithmetical progression. Thus the progression is given by

$$a, a + r, a + 2r, \ldots .$$

One of the terms is

$$a + ar = a(1 + r).$$

Another one is

$$a + (2a + ar)r = a(1 + r)^2.$$

It is not difficult to see that all numbers of the form $a(1 + r)^n$, $n \geq 1$ are terms of a geometrical progression.

Problem 3.91 The set of positive integers is partitioned into several arithmetical progressions. Prove that in at least one of them the first term is divisible by the common difference.

Solution Let $a_i, r_i, i = 1, 2, \ldots, n$ be the first terms and the common differences of the progressions. Consider the number

$$r = r_1 r_2 \ldots r_n.$$

This number must belong to one of the progressions, thus, for some i and k we have

$$r_1 r_2 \ldots r_n = a_i + k r_i.$$

It follows that a_i is divisible by r_i.

Try to solve the following problems.

Problem 3.92 Partition the set of positive integers into two subsets such that neither of them contains a non-constant arithmetical progression.

Problem 3.93 Prove that among the terms of the progression $3, 7, 11, \ldots$ there are infinitely many prime numbers.

Problem 3.94 Does there exist an (infinite) non-constant arithmetical progression whose terms are all prime numbers?

Problem 3.95 Consider an arithmetical progression of positive integers. Prove that one can find infinitely many terms the sum of whose decimal digits is the same.

Problem 3.96 The set of positive integers is partitioned into n arithmetical progressions, with common differences $r_1, r_2, \ldots, r_n$. Prove that

$$\frac{1}{r_1} + \frac{1}{r_2} + \cdots + \frac{1}{r_n} = 1.$$

Problem 3.97 Prove that for every positive integer n one can find n integers in arithmetical progression, all of them nontrivial powers of some integers, but one cannot find an infinite sequence with this property.

Problem 3.98 Prove that for any integer n, $n \geq 3$, there exist n positive integers in arithmetical progression $a_1, a_2, \ldots, a_n$ and n positive integers in geometric progression $b_1, b_2, \ldots, b_n$, such that

$$b_1 < a_1 < b_2 < a_2 < \cdots < b_n < a_n.$$

Problem 3.99 Let $(a_n)_{n \geq 1}$ be an arithmetic sequence such that a_1^2, a_2^2, and a_3^2 are also terms of the sequence. Prove that the terms of this sequence are all integers.

Problem 3.100 Let $A = \{1, \frac{1}{2}, \frac{1}{3}, \frac{1}{4}, \ldots\}$. Prove that for every positive integer $n \geq 3$ the set A contains a non-constant arithmetic sequence of length n, but it does not contain an infinite non-constant arithmetic sequence.

Problem 3.101 Let n be a positive integer and let $x_1 \leq x_2 \leq \cdots \leq x_n$ be real numbers. Prove that

$$\left(\sum_{i,j=1}^{n} |x_i - x_j| \right)^2 \leq \frac{2(n^2 - 1)}{3} \sum_{i,j=1}^{n} (x_i - x_j)^2.$$

Show that the equality holds if and only if $x_1, \ldots, x_n$ is an arithmetic sequence.

3.9 The Marriage Lemma

A number of n boys and n girls are attending a party. Each boy is acquainted with some of the girls. Under which conditions is it possible that at some dance each of the boys dances with a girl of his acquaintance? (We will call this a matching between the boys and the girls.)

Clearly, first of all, each boy must be acquainted with at least one girl. This is not sufficient, though, since there might exist two boys acquainted with only one girl, the same for both of them. Therefore we derive another necessary condition: for any pair of boys, the set of girls acquainted with at least one of them must have at least two elements. Furthermore, for any set of three boys, the set of girls acquainted with at least one of them must have at least three elements.

Let us state now what seems to be a reasonable necessary condition: there exists a matching between the two sets if and only if for every k, $1 \le k \le n$, and every set of k boys, the set of girls acquainted with at least one of them must have at least k elements.

It turns out that this condition is also a sufficient one, as proved by the British mathematician Philip Hall in 1935. We will call this Hall's condition.

The proof goes by induction on n. If $n = 1$ there is nothing to prove. Assume the assertion true for all $k \le n$ and consider a set of $n + 1$ boys and $n + 1$ girls satisfying Hall's condition. We distinguish two cases.

Case 1. If any $k \le n$ boys are acquainted with at least $k + 1$ girls then choose one of the boys and let him dance with any of the girls of his acquaintance. Clearly, the Hall's condition is fulfilled by the remaining boys and girls and the induction hypothesis applies.

Case 2. If there exist a set B of $k \le n$ boys who are acquainted with exactly k girls, then these boys can be paired with those girls. We claim that Hall's condition is still fulfilled by the remaining boys and girls, thus they can also be paired. Indeed, let X be set of $x \le n - k + 1$ boys. If the set of girls acquainted to these boys has less than x elements, then the set of girls acquainted to the boys in the set $X \cup B$ has less than $x + k$ elements. But the set $X \cup B$ has exactly $x + k$ elements. This contradicts the assumption that Hall's condition is fulfilled by the initial set of boys and girls.

The lemma still works if the number of girls is greater than the number of boys. Just add a necessary number of boys who know all girls to even them up.

Let us apply the lemma in solving some problems.

Problem 3.102 Two square sheets have areas equal to 2003. Each of the sheets is arbitrarily divided into 2003 non-overlapping polygons, besides, each of the polygons has an unitary area. Afterward, one overlays the sheets, and it is asked to prove that the obtained double layer can be punctured 2003 times, so that each of the 4006 polygons gets punctured precisely once.

Solution Consider the polygons painted blue on one sheet and red on the other one. The boys will be the blue polygons while the girls the red ones. Now, a boy is acquainted to a girl if the blue and red polygons overlap. Clearly, it suffices to check that Hall's condition is fulfilled to obtain the desired conclusion.

Suppose, by way of contradiction, that Hall's condition is violated. Then we can find k blue polygons such that the corresponding red ones are at most $k - 1$. This leads to a contradiction if we look at the areas. The area of the union of the blue k polygons equals k, and this cannot be covered with less than k red polygons.

Problem 3.103 Some pieces are placed on an 8×8 table. There are exactly 4 pieces in each row and each column of the board. Show that there are 8 pieces among those pieces such that no two of them are in the same row or column.

Solution Let us examine a possible example of such a table:

	g_1	g_2	g_3	g_4	g_5	g_6	g_7	g_8
b_1	□			■		□		□
b_2		□	■		□		□	
b_3			□		■	□		□
b_4	■	□			□		□	
b_5			■	□	□		□	
b_6	□			□		□		■
b_7	□		□		□		■	
b_8		□		□		■		□

Surely, the alert reader can see (looking at the notations) who are the boys and girls.

Indeed, the boys are the rows of the table, while the girls are the columns. A boy b_i is acquainted with the girl g_j iff there is a piece at the intersection of the ith row with the jth column. For instance, in the previous example, the first boy b_1 is acquainted with the girls g_1, g_4, g_6, and g_8, etc. All we have to do is to prove that there exists a matching between the rows and the columns, hence we must show that Hall's condition holds.

Consider a set of k rows, $1 \le k \le 8$, and observe that they are acquainted with $4k$ (not necessarily distinct) columns, since in each row there are 4 pieces. But a column may appear at most 4 times, since in each column there are 4 pieces, as well. Therefore, the number of distinct columns corresponding to the chosen k rows is at least $4k/4 = k$, hence Hall's condition is fulfilled.

Observation A similar argument can be used to prove a more general statement. If every boy is acquainted to exactly m girls, and every girl is acquainted with exactly m boys, then Hall's condition is fulfilled and there is a matching between the boys and the girls.

Now, find the boys and the girls in the following proposed problems.

Problem 3.104 A deck of cards is arranged, face up, in a 4×13 array. Prove that one can pick a card from each column in such a way as to get one card of each denomination.

Problem 3.105 An $n \times n$ table is filled with 0 and 1 so that if we chose randomly n cells (no two of them in the same row or column) then at least one contains 1. Prove that we can find i rows and j columns so that $i + j \geq n + 1$ and their intersection contains only 1's.

Problem 3.106 Let X be a finite set and let $\bigsqcup_{i=1}^{n} X_i = \bigsqcup_{j=1}^{n} Y_j$ be two disjoint decompositions with all sets X_i's and Y_j's having the same size. Prove that there exist distinct elements $x_1, x_2, \ldots, x_n$ which are in different sets in both decompositions.

Problem 3.107 A set P consists of 2005 distinct prime numbers. Let A be the set of all possible products of 1002 elements of P, and B be the set of all products of 1003 elements of P. Prove the existence of a one-to-one correspondence f from A to B with the property that a divides $f(a)$ for all $a \in A$.

Problem 3.108 The entries of a $n \times n$ table are non-negative real numbers such that the numbers in each row and column add up to 1. Prove that one can pick n numbers from distinct rows and columns which are positive.

Problem 3.109 There are b boys and g girls present at a party, where b and g are positive integers satisfying $g \geq 2b - 1$. Each boy invites a girl for a dance (of course, two different boys must always invite two different girls). Prove that this can be done in such a way that every boy is either dancing with a girl he knows or all the girls he knows are not dancing.

Problem 3.110 A $m \times n$ array is filled with the numbers $1, 2, \ldots, n$, each used exactly m times. Show that one can always permute the numbers within columns to arrange that each row contains every number $1, 2, \ldots, n$ exactly once.

Problem 3.111 Some of the AwesomeMath students went on a trip to the beach. There were provided n buses of equal capacities for both the trip to the beach and the ride home, one student in each seat, and there were not enough seats in $n - 1$ buses to fit each student. Every student who left in a bus came back in a bus, but not necessarily the same one.

Prove that there are n students such that any two were on different busses on both rides.

Part II
Solutions

Chapter 4
Algebra

4.1 An Algebraic Identity

Problem 1.4 Factor $(a + 2b - 3c)^3 + (b + 2c - 3a)^3 + (c + 2a - 3b)^3$.

Solution Observe that $(a + 2b - 3c) + (b + 2c - 3a) + (c + 2a - 3b) = 0$. Because $x + y + z = 0$ implies $x^3 + y^3 + z^3 = 3xyz$, we obtain

$$(a + 2b - 3c)^3 + (b + 2c - 3a)^3 + (c + 2a - 3b)^3$$
$$= 3(a + 2b - 3c)(b + 2c - 3a)(c + 2a - 3b).$$

Problem 1.5 Let x, y, z be integers such that

$$(x - y)^2 + (y - z)^2 + (z - x)^2 = xyz.$$

Prove that $x^3 + y^3 + z^3$ is divisible by $x + y + z + 6$.

Solution We first notice that at least one of x, y, z is even. Indeed, if x, y, z are all odd, the left-hand side of the given equality is even while the right-hand side is odd, a contradiction. Because

$$x^3 + y^3 + z^3 - 3xyz = (x + y + z)(x^2 + y^2 + z^2 - xy - yz - zx),$$

and

$$x^2 + y^2 + z^2 - xy - yz - zx$$
$$= \frac{1}{2}\left((x - y)^2 + (y - z)^2 + (z - x)^2\right) = \frac{xyz}{2}$$

we obtain

$$x^3 + y^3 + z^3 - 3xyz = (x + y + z)\frac{xyz}{2},$$

T. Andreescu, B. Enescu, *Mathematical Olympiad Treasures*,
DOI 10.1007/978-0-8176-8253-8_4, © Springer Science+Business Media, LLC 2011

hence

$$x^3 + y^3 + z^3 = \frac{xyz}{2}(x + y + z + 6),$$

and since $\frac{xyz}{2}$ is an integer, the conclusion follows.

Problem 1.6 Let a, b, c be distinct real numbers. Prove that the following equality cannot hold:

$$\sqrt[3]{a - b} + \sqrt[3]{b - c} + \sqrt[3]{c - a} = 0$$

Solution Suppose $\sqrt[3]{a - b} + \sqrt[3]{b - c} + \sqrt[3]{c - a} = 0$. Then

$$(a - b) + (b - c) + (c - a) = 3\sqrt[3]{(a - b)(b - c)(c - a)}.$$

This implies

$$(a - b)(b - c)(c - a) = 0,$$

which contradicts the hypothesis that a, b, c are distinct.

Problem 1.7 Prove that the number

$$\sqrt[3]{45 + 29\sqrt{2}} + \sqrt[3]{45 - 29\sqrt{2}}$$

is a rational number.

Solution Let $a = \sqrt[3]{45 + 29\sqrt{2}} + \sqrt[3]{45 - 29\sqrt{2}}$. Because

$$a - \sqrt[3]{45 + 29\sqrt{2}} - \sqrt[3]{45 - 29\sqrt{2}} = 0,$$

we have

$$a^3 - \left(45 + 29\sqrt{2}\right) - \left(45 - 29\sqrt{2}\right) = 3a\sqrt[3]{\left(45 + 29\sqrt{2}\right)\left(45 - 29\sqrt{2}\right)},$$

which is equivalent to

$$a^3 - 21a - 90 = 0,$$

or

$$(a - 6)\left(a^2 + 6a + 15\right) = 0.$$

The equation $a^2 + 6a + 15 = 0$ has no real roots; hence

$$\sqrt[3]{45 + 29\sqrt{2}} + \sqrt[3]{45 - 29\sqrt{2}} = 6,$$

a rational number.

Problem 1.8 Let a, b, c be rational numbers such that

$$a + b\sqrt[3]{2} + c\sqrt[3]{4} = 0.$$

Prove that $a = b = c = 0$.

Solution Notice first that we can assume that a, b, c are integers, otherwise we can multiply the equality by the least common multiple of the denominators. Next, if a, b, c are not equal to 0, we can assume that they are not all even; if they were, we could simplify the equation by dividing both sides by the greatest common power of two.

We have seen that $x + y + z = 0$ implies

$$x^3 + y^3 + z^3 = 3xyz.$$

Taking $x = a$, $y = b\sqrt[3]{2}$ and $z = c\sqrt[3]{4}$ yields

$$a^3 + 2b^3 + 4c^3 = 6abc,$$

thus a must be even, say $a = 2a_1$. Plugging this in the above equality yields

$$8a_1^3 + 2b^3 + 4c^3 = 12a_1bc,$$

or

$$4a_1^3 + b^3 + 2c^3 = 6a_1bc,$$

hence b is also even.

By similar arguments, c is also even; which is a contradiction.

Problem 1.9 Let r be a real number such that

$$\sqrt[3]{r} + \frac{1}{\sqrt[3]{r}} = 3.$$

Determine the value of

$$r^3 + \frac{1}{r^3}.$$

Solution From $\sqrt[3]{r} + \frac{1}{\sqrt[3]{r}} - 3 = 0$, we obtain

$$r + \frac{1}{r} - 27 = 3\sqrt[3]{r}\frac{1}{\sqrt[3]{r}}(-3) = -9.$$

Therefore,

$$r + \frac{1}{r} - 18 = 0.$$

It follows that

$$r^3 + \frac{1}{r^3} - 18^3 = 3r\frac{1}{r}(-18) = -54.$$

Finally,

$$r^3 + \frac{1}{r^3} = 18^3 - 54 = 5778.$$

Problem 1.10 Find the locus of points (x, y) for which

$$x^3 + y^3 + 3xy = 1.$$

Solution Observe that the given equality is equivalent to

$$x^3 + y^3 + (-1)^3 - 3xy(-1) = 0,$$

which can be written as

$$(x + y - 1)(x^2 + y^2 - xy + x + y + 1) = 0.$$

Note that we can write the second factor more conveniently:

$$x^2 + y^2 - xy + x + y + 1 = \frac{1}{2}[(x - y)^2 + (x + 1)^2 + (y + 1)^2],$$

it follows that the locus consists of the line $x + y - 1 = 0$ and the point $(-1, -1)$.

Problem 1.11 Let n be a positive integer. Prove that the number

$$3^{3^n}\left(3^{3^n} + 1\right) + 3^{3^n + 1} - 1,$$

is not a prime.

Solution Observe that

$$3^{3^n}\left(3^{3^n} + 1\right) + 3^{3^n + 1} - 1 = a^3 + b^3 + c^3 - 3abc,$$

where $a = 3^{3^{n-1}}$, $b = 9^{3^{n-1}}$ and $c = -1$. Thus, the number factors as

$$\left(3^{3^{n-1}} + 9^{3^{n-1}} - 1\right)\left(9^{3^{n-1}} + 81^{3^{n-1}} + 1 - 27^{3^{n-1}} + 3^{3^{n-1}} + 9^{3^{n-1}}\right).$$

Problem 1.12 Let S be the set of integers x such that $x = a^3 + b^3 + c^3 - 3abc$, for some integers a, b, c. Prove that if $x, y \in S$, then $xy \in S$.

Solution We have seen that

$$a^3 + b^3 + c^3 - 3abc = (a + b + c)(a + b\omega + c\omega^2)(a + b\omega^2 + c\omega),$$

where $\omega = \frac{-1+i\sqrt{3}}{2}$ is one of the primitive cubic roots of unity. Let the polynomial $P(X) = a + bX + cX^2$. We obtain

$$a^3 + b^3 + c^3 - 3abc = P(1)P(\omega)P(\omega^2).$$

We observe that $x \in S$ if and only if there exists a polynomial P with integer coefficients and degree at most 2 such that

$$x = P(1)P(\omega)P(\omega^2).$$

Now, let $x, y \in S$, and let P and Q be polynomials with integer coefficients, of degree at most 2 such that $x = P(1)P(\omega)P(\omega^2)$ and $y = Q(1)Q(\omega)Q(\omega^2)$. Then $xy = R(1)R(\omega)R(\omega^2)$, where $R(X) = P(X)Q(X)$ is a polynomial of degree at most 4. If we divide R by $X^3 - 1$, we obtain a remainder $R_1(X)$, with integer coefficients and degree at most 2 (this follows from the division algorithm). Thus

$$R(X) = (X^3 - 1)C(X) + R_1(X)$$

for some polynomial $C(X)$.

But then $R(1) = R_1(1)$, $R(\omega) = R_1(\omega)$, and $R(\omega^2) = R_1(\omega^2)$, since 1, ω and ω^2 are the roots of $X^3 - 1$.

It follows that $xy = R_1(1)R_1(\omega)R_1(\omega^2)$, where R_1 is a polynomial with integer coefficients, of degree at most 2, hence $xy \in S$.

Observation Another proof can be given if we recall that

$$a^3 + b^3 + c^3 - 3abc = \begin{vmatrix} a & b & c \\ c & a & b \\ b & c & a \end{vmatrix}.$$

Let

$$x = \begin{vmatrix} a & b & c \\ c & a & b \\ b & c & a \end{vmatrix} \quad \text{and} \quad y = \begin{vmatrix} a' & b' & c' \\ c' & a' & b' \\ b' & c' & a' \end{vmatrix},$$

where a, b, c, a', b', c' are integers. Then

$$xy = \begin{vmatrix} aa' + bc' + cb' & ab' + ba' + cc' & ac' + bb' + ca' \\ ac' + bb' + ca' & aa' + bc' + cb' & ab' + ba' + cc' \\ ab' + ba' + cc' & ac' + bb' + ca' & aa' + bc' + cb' \end{vmatrix} \in S.$$

Problem 1.13 Let a, b, c be distinct positive integers and let k be a positive integer such that

$$ab + bc + ca \geq 3k^2 - 1.$$

Prove that

$$\frac{a^3 + b^3 + c^3}{3} - abc \geq 3k.$$

Solution The desired inequality is equivalent to

$$a^3 + b^3 + c^3 - 3abc \geq 9k.$$

Assume, with no loss of generality, that $a > b > c$. Then $a - b \geq 1$, $b - c \geq 1$ and $a - c \geq 2$. It follows that

$$a^2 + b^2 + c^2 - ab - bc - ac$$
$$= \frac{1}{2}\left((a-b)^2 + (b-c)^2 + (a-c)^2\right) \geq \frac{1}{2}(1 + 1 + 4) = 3.$$

We obtain

$$a^3 + b^3 + c^3 - 3abc = (a+b+c)\left(a^2 + b^2 + c^2 - ab - bc - ac\right)$$
$$\geq 3(a+b+c),$$

so it suffices to prove that

$$3(a+b+c) \geq 9k,$$

or

$$a + b + c \geq 3k.$$

But

$$(a+b+c)^2 = a^2 + b^2 + c^2 + 2ab + 2bc + 2ac$$
$$= a^2 + b^2 + c^2 - ab - bc - ac + 3(ab + bc + ac)$$
$$\geq 3 + 3(3k^2 - 1) = 9k^2,$$

and the conclusion follows.

Problem 1.14 Let a, b, and c be the side lengths of a triangle. Prove that

$$\sqrt[3]{\frac{a^3 + b^3 + c^3 + 3abc}{2}} \geq \max(a, b, c).$$

Solution Assume, with no loss of generality, that $a \geq b \geq c$. We have to prove that

$$\sqrt[3]{\frac{a^3 + b^3 + c^3 + 3abc}{2}} \geq a,$$

which is equivalent to

$$-a^3 + b^3 + c^3 + 3abc \geq 0.$$

Since

$$-a^3 + b^3 + c^3 + 3abc = (-a)^3 + b^3 + c^3 - 3(-a)bc,$$

the latter expression factors into

$$\frac{1}{2}(-a + b + c)\big((a + b)^2 + (a + c)^2 + (b - c)^2\big).$$

The conclusion now follows from the triangle inequality: namely, $b + c > a$ in any triangle.

Problem 1.15 Find the least real number r such that for each triangle with side lengths a, b, c,

$$\frac{\max(a, b, c)}{\sqrt[3]{a^3 + b^3 + c^3 + 3abc}} < r.$$

Solution Plugging $a = b = n > 1$ and $c = 1$ yields

$$\frac{n}{\sqrt[3]{2n^3 + 3n^2 + 1}} < r.$$

Since

$$\lim_{n \to \infty} \frac{n}{\sqrt[3]{2n^3 + 3n^2 + 1}} = \lim_{n \to \infty} \frac{1}{\sqrt[3]{2 + \frac{3}{n} + \frac{1}{n^3}}} = \frac{1}{\sqrt[3]{2}},$$

it follows that $r \geq \frac{1}{\sqrt[3]{2}}$. We will show that

$$\frac{\max(a, b, c)}{\sqrt[3]{a^3 + b^3 + c^3 + 3abc}} < \frac{1}{\sqrt[3]{2}}$$

holds for each triangle. Indeed, with no loss of generality we can assume that $a = \max(a, b, c)$. Then, the latter is equivalent to

$$-a^3 + b^3 + c^3 + 3abc > 0,$$

which factors as

$$(-a + b + c)\big((a + b)^2 + (a + c)^2 + (b - c)^2\big) > 0,$$

obviously true.

Problem 1.16 Find all integers that can be represented as $a^3 + b^3 + c^3 - 3abc$ for some positive integers $a, b,$ and c.

Solution Call an integer nice if it can be represented as $a^3 + b^3 + c^3 - 3abc$ for some positive integers a, b, c. Assume without loss of generality that $b = a + x$ and $c = a + x + y$, for some nonnegative integers x, y. Therefore,

$$a^3 + b^3 + c^3 - 3abc = (3a + 2x + y)(x^2 + xy + y^2).$$

For $x = y = 0$ it follows that 0 is nice. If x and y are not both 0, then

$$(3a + 2x + y)(x^2 + xy + y^2) \geq 3a + 1,$$

hence $1, 2$ and 3 are not nice. We claim that if a nice integer is divisible by 3, then it is divisible by 9. Indeed, we have

$$0 \equiv (3a + 2x + y)(x^2 + xy + y^2) \equiv (y - x) \cdot (x - y)^2 \equiv (y - x)^3 \pmod{3},$$

hence $x \equiv y \pmod{3}$. Therefore

$$3a + 2x + y \equiv x^2 + xy + y^2 \equiv 0 \pmod{3}$$

and it follows that

$$(3a + 2x + y)(x^2 + xy + y^2) \equiv 0 \pmod{9}.$$

However, 9 itself is not nice, since

$$(3a + 2x + y)(x^2 + xy + y^2) \geq (3a + 3)3 > 9.$$

Finally, let us proceed to find which integers are nice. Taking $x = 0$, $y = 1$ it follows that any positive integer of the form $3a + 1$ is nice. Taking $x = 1$, $y = 0$ it follows that any positive integer of the form $3a + 2$ is nice. Taking $x = y = 1$ it follows that any positive integer of the form $9(a + 1)$ is nice. From these we conclude that all the nice integers are 0, any positive integer greater than 3 of the form $3a + 1$ or $3a + 2$, and the integers greater than 9 of the form $9a$.

Problem 1.17 Find all pairs (x, y) of integers such that

$$xy + \frac{x^3 + y^3}{3} = 2007.$$

Solution Rewrite the equation as

$$x^3 + y^3 - 1 + 3xy = 6020.$$

Factoring both sides yields

$$(x + y - 1)(x^2 + y^2 + 1 + x + y - xy) = 2^2 \cdot 5 \cdot 7 \cdot 43.$$

Since $x^2 + y^2 + 1 + x + y - xy = (x + y - 1)^2 + 3x + 3y - 3xy$, taking the above equality mod 3, we deduce that

$$x + y - 1 \equiv 2 \pmod 3.$$

Also, $x + y - 1$ is less than $x^2 + y^2 + 1 + x + y - xy$, so the possible candidates for $x + y - 1$ are 2, 5, 14, 20 and 35.

By inspection, we obtain $x + y - 1 = 20$ and, finally, $(x, y) = (18, 3)$ or $(x, y) = (3, 18)$.

Problem 1.18 Let k be an integer and let

$$n = \sqrt[3]{k + \sqrt{k^2 - 1}} + \sqrt[3]{k - \sqrt{k^2 - 1}} + 1.$$

Prove that $n^3 - 3n^2$ is an integer.

Solution Let $a = \sqrt[3]{k + \sqrt{k^2 - 1}}$, $b = \sqrt[3]{k - \sqrt{k^2 - 1}}$, and $c = 1 - n$. Then $a + b + c = 0$, therefore

$$a^3 + b^3 + c^3 = 3abc.$$

Since $ab = 1$, this is equivalent to

$$2k + (1 - n)^3 = 3(1 - n),$$

or

$$n^3 - 3n^2 = 2k - 2,$$

obviously an integer number.

4.2 Cauchy–Schwarz Revisited

Problem 1.21 Let $x, y, z > 0$. Prove that

$$\frac{2}{x + y} + \frac{2}{y + z} + \frac{2}{z + x} \geq \frac{9}{x + y + z}.$$

Solution Writing the left-hand side as

$$\frac{(\sqrt{2})^2}{x + y} + \frac{(\sqrt{2})^2}{y + z} + \frac{(\sqrt{2})^2}{z + x},$$

and using the lemma in p. 8, we deduce that

$$\frac{(\sqrt{2})^2}{x + y} + \frac{(\sqrt{2})^2}{y + z} + \frac{(\sqrt{2})^2}{z + x} \geq \frac{(3\sqrt{2})^2}{2(x + y + z)} = \frac{9}{x + y + z}.$$

Equality occurs when $x = y = z$.

Problem 1.22 Let a, b, x, y, z be positive real numbers. Prove that

$$\frac{x}{ay + bz} + \frac{y}{az + bx} + \frac{z}{ax + by} \geq \frac{3}{a + b}.$$

Solution We have

$$\frac{x}{ay + bz} + \frac{y}{az + bx} + \frac{z}{ax + by} = \frac{x^2}{axy + bxz} + \frac{y^2}{ayz + bxy} + \frac{z^2}{axz + byz}$$

$$\geq \frac{(x + y + z)^2}{(a + b)(xy + xz + yz)} \geq \frac{3}{a + b}.$$

Problem 1.23 Let $a, b, c > 0$. Prove that

$$\frac{a^2 + b^2}{a + b} + \frac{b^2 + c^2}{b + c} + \frac{a^2 + c^2}{a + c} \geq a + b + c.$$

Solution We write the left-hand side as

$$\frac{a^2}{a + b} + \frac{b^2}{b + c} + \frac{c^2}{a + c} + \frac{b^2}{a + b} + \frac{c^2}{b + c} + \frac{a^2}{a + c}.$$

Applying the same lemma, we find that this expression is greater than or equal to

$$\frac{(2a + 2b + 2c)^2}{4(a + b + c)} = a + b + c.$$

Problem 1.24 Let a, b, c be positive numbers such that $abc = 1$. Prove that

$$\frac{1}{a^3(b + c)} + \frac{1}{b^3(a + c)} + \frac{1}{c^3(a + b)} \geq \frac{3}{2}.$$

Solution We see that

$$\frac{1}{a^3(b + c)} + \frac{1}{b^3(a + c)} + \frac{1}{c^3(a + b)} = \frac{\frac{1}{a^2}}{ab + ac} + \frac{\frac{1}{b^2}}{ab + bc} + \frac{\frac{1}{c^2}}{ac + bc}$$

$$\geq \frac{(\frac{1}{a} + \frac{1}{b} + \frac{1}{c})^2}{2(ab + bc + ac)}.$$

The last inequality comes from the lemma.
    Because $abc = 1$, it follows that

$$\left(\frac{1}{a} + \frac{1}{b} + \frac{1}{c}\right)^2 = \frac{(ab + bc + ca)^2}{(abc)^2} = (ab + bc + ca)^2.$$

Hence

$$\frac{1}{a^3(b + c)} + \frac{1}{b^3(a + c)} + \frac{1}{c^3(a + b)} \geq \frac{(ab + bc + ca)}{2} \geq \frac{3\sqrt[3]{(abc)^2}}{2} = \frac{3}{2}.$$

Problem 1.25 Let $x, y, z > 0$. Prove that

$$\frac{x}{x + 2y + 3z} + \frac{y}{y + 2z + 3x} + \frac{z}{z + 2x + 3y} \geq \frac{1}{2}.$$

Solution We write the left-hand side as

$$\frac{x^2}{x^2 + 2xy + 3xz} + \frac{y^2}{y^2 + 2yz + 3xy} + \frac{z^2}{z^2 + 2xz + 3yz}$$

and apply the lemma. This yields

$$\frac{x}{x + 2y + 3z} + \frac{y}{y + 2z + 3x} + \frac{z}{z + 2x + 3y} \geq \frac{(x + y + z)^2}{x^2 + y^2 + z^2 + 5(xy + xz + yz)}.$$

Now it suffices to prove that

$$\frac{(x + y + z)^2}{x^2 + y^2 + z^2 + 5(xy + xz + yz)} \geq \frac{1}{2},$$

but this is equivalent to

$$x^2 + y^2 + z^2 \geq xy + xz + yz,$$

proved in Problem 1.19.

Problem 1.26 Let $x, y, z > 0$. Prove that

$$\frac{x^2}{(x + y)(x + z)} + \frac{y^2}{(y + z)(y + x)} + \frac{z^2}{(z + x)(z + y)} \geq \frac{3}{4}.$$

Solution We apply the lemma again to obtain

$$\frac{x^2}{(x + y)(x + z)} + \frac{y^2}{(y + z)(y + x)} + \frac{z^2}{(z + x)(z + y)}$$

$$\geq \frac{(x + y + z)^2}{x^2 + y^2 + z^2 + 3(xy + xz + yz)}.$$

The inequality

$$\frac{(x + y + z)^2}{x^2 + y^2 + z^2 + 3(xy + xz + yz)} \geq \frac{3}{4}$$

is equivalent to

$$x^2 + y^2 + z^2 \geq xy + xz + yz.$$

Problem 1.27 Let a, b, c, d, e be positive real numbers. Prove that

$$\frac{a}{b+c} + \frac{b}{c+d} + \frac{c}{d+e} + \frac{d}{e+a} + \frac{e}{a+b} \geq \frac{5}{2}.$$

Solution We have

$$\frac{a}{b+c} + \frac{b}{c+d} + \frac{c}{d+e} + \frac{d}{e+a} + \frac{e}{a+b}$$

$$= \frac{a^2}{ab+ac} + \frac{b^2}{bc+bd} + \frac{c^2}{cd+ce} + \frac{d^2}{de+ad} + \frac{e^2}{ae+be}$$

$$\geq \frac{(a+b+c+d+e)^2}{\sum ab}.$$

Because

$$(a+b+c+d+e)^2 = \sum a^2 + 2\sum ab,$$

we have to prove that

$$2\sum a^2 + 4\sum ab \geq 5\sum ab,$$

which is equivalent to

$$2\sum a^2 \geq \sum ab.$$

The last inequality follows from $\sum (a-b)^2 \geq 0$.

Problem 1.28 Let a, b, c be positive real numbers such that

$$ab + bc + ca = \frac{1}{3}.$$

Prove that

$$\frac{a}{a^2 - bc + 1} + \frac{b}{b^2 - ca + 1} + \frac{c}{c^2 - ab + 1} \geq \frac{1}{a+b+c}.$$

Solution We apply our lemma:

$$\sum \frac{a}{a^2 - bc + 1} = \sum \frac{a^2}{a^3 - abc + a} \geq \frac{(a+b+c)^2}{\sum a^3 + \sum a - 3abc}.$$

Because

$$\sum a^3 - 3abc = (a+b+c)\left(\sum a^2 - \sum ab\right) = (a+b+c)\left(\sum a^2 - \frac{1}{3}\right)$$

it follows that

$$\frac{(a+b+c)^2}{\sum a^3 + \sum a - 3abc} = \frac{\sum a^2 + 2\sum ab}{(a+b+c)(\sum a^2 - \frac{1}{3} + 1)}$$

$$= \frac{\sum a^2 + \frac{2}{3}}{(a+b+c)(\sum a^2 + \frac{2}{3})} = \frac{1}{a+b+c},$$

as desired.

Problem 1.29 Let a, b, c be positive real numbers such that $abc = 1$. Prove that

$$\frac{a+b+1}{a+b^2+c^3} + \frac{b+c+1}{b+c^2+a^3} + \frac{c+a+1}{c+a^2+b^3} \le \frac{(a+1)(b+1)(c+1)+1}{a+b+c}.$$

Solution We have

$$a+b^2+c^3 = \frac{a^2}{a} + \frac{b^2}{1} + \frac{c^2}{ab} \ge \frac{(a+b+c)^2}{a+1+ab},$$

thus

$$\frac{a+b+1}{a+b^2+c^3} \le \frac{(a+b+1)(a+1+ab)}{(a+b+c)^2}.$$

Now, write the similar two inequalities and add them all up. We only have to check that

$$\sum (a+b+1)(a+1+ab) = (a+b+c)\big((a+1)(b+1)(c+1)+1\big),$$

which is trivial.

Problem 1.30 Let a and b be positive real numbers. Prove that

$$\frac{a^3+b^3}{a^4+b^4} \cdot \frac{a+b}{a^2+b^2} \ge \frac{a^4+b^4}{a^6+b^6}.$$

Solution The inequality rewrites as

$$(a+b)(a^3+b^3)(a^6+b^6) \ge (a^2+b^2)(a^4+b^4)^2.$$

Observe that

$$a^3+b^3 = \frac{a^4}{a} + \frac{b^4}{b} \ge \frac{(a^2+b^2)^2}{a+b},$$

hence

$$(a+b)(a^3+b^3) \ge (a^2+b^2)^2.$$

Similarly,

$$a^6 + b^6 = \frac{a^8}{a^2} + \frac{b^8}{b^2} \geq \frac{(a^4 + b^4)^2}{a^2 + b^2}.$$

Multiplying out the last two inequalities yields the result.

Problem 1.31 Let a, b, c be positive real numbers such that $ab + bc + ca \geq 3$. Prove that

$$\frac{a}{\sqrt{a+b}} + \frac{b}{\sqrt{b+c}} + \frac{c}{\sqrt{c+a}} \geq \frac{3}{\sqrt{2}}.$$

Solution We have

$$\frac{a}{\sqrt{a+b}} + \frac{b}{\sqrt{b+c}} + \frac{c}{\sqrt{c+a}} = \frac{a^2}{a\sqrt{a+b}} + \frac{b^2}{b\sqrt{b+c}} + \frac{c^2}{c\sqrt{c+a}},$$

so our lemma yields

$$\frac{a}{\sqrt{a+b}} + \frac{b}{\sqrt{b+c}} + \frac{c}{\sqrt{c+a}} \geq \frac{(a+b+c)^2}{a\sqrt{a+b} + b\sqrt{b+c} + c\sqrt{c+a}}.$$

On the other hand, applying again the lemma,

$$\sum a(a+b) = \sum \frac{a^2(a+b)}{a} \geq \frac{(a\sqrt{a+b} + b\sqrt{b+c} + c\sqrt{c+a})^2}{a+b+c},$$

hence

$$\frac{(a+b+c)^2}{a\sqrt{a+b} + b\sqrt{b+c} + c\sqrt{c+a}} \geq \frac{(a+b+c)^{\frac{3}{2}}}{\sqrt{a^2 + b^2 + c^2 + ab + bc + ca}}.$$

The inequality

$$\frac{(a+b+c)^{\frac{3}{2}}}{\sqrt{a^2 + b^2 + c^2 + ab + bc + ca}} \geq \frac{3}{\sqrt{2}}$$

is equivalent to

$$2(a+b+c)^3 \geq 9(a^2 + b^2 + c^2 + ab + bc + ca),$$

which can be written as

$$2(a+b+c)^3 - 9(a+b+c)^2 + 9(ab + bc + ca) \geq 0.$$

Since $ab + bc + ca \geq 3$, it suffices to show that

$$2(a+b+c)^3 - 9(a+b+c)^2 + 27 \geq 0.$$

Denote $a + b + c = s$. Then

$$2s^3 - 9s^2 + 27 = (s - 3)^2(2s + 3) \geq 0,$$

as desired.

Problem 1.32 Let a, b, c be positive real numbers. Prove that

$$\frac{a}{b(b+c)^2} + \frac{b}{c(c+a)^2} + \frac{c}{a(a+b)^2} \geq \frac{9}{4(ab+bc+ca)}.$$

Solution Observe that

$$\frac{a}{b(b+c)^2} = \frac{\frac{a^2}{(b+c)^2}}{ab}.$$

Writing the similar two equalities and using the lemma yield

$$\frac{a}{b(b+c)^2} + \frac{b}{c(c+a)^2} + \frac{c}{a(a+b)^2} \geq \frac{(\frac{a}{b+c} + \frac{b}{c+a} + \frac{c}{a+b})^2}{ab+bc+ca}.$$

We know (see Problem 1.19 of Sect. 1.2) that

$$\frac{a}{b+c} + \frac{b}{c+a} + \frac{c}{a+b} \geq \frac{3}{2},$$

hence

$$\frac{(\frac{a}{b+c} + \frac{b}{c+a} + \frac{c}{a+b})^2}{ab+bc+ca} \geq \frac{9}{4(ab+bc+ca)}.$$

Problem 1.33 Let a, b, c be positive real numbers such that

$$\frac{1}{a^2+b^2+1} + \frac{1}{b^2+c^2+1} + \frac{1}{c^2+a^2+1} \geq 1.$$

Prove that

$$ab + bc + ca \leq 3.$$

Solution We have

$$a^2 + b^2 + 1 = \frac{a^2}{1} + \frac{b^2}{1} + \frac{c^2}{c^2} \geq \frac{(a+b+c)^2}{2+c^2},$$

so

$$\frac{1}{a^2+b^2+1} \leq \frac{2+c^2}{(a+b+c)^2}.$$

The similar two inequalities and the hypothesis imply

$$\frac{6 + a^2 + b^2 + c^2}{(a + b + c)^2} \geq 1,$$

which quickly simplifies to

$$ab + bc + ca \leq 3.$$

4.3 Easy Ways Through Absolute Values

Problem 1.39 Solve the equation $|x - 3| + |x + 1| = 4$.

Solution Using the inequality $|a| + |b| \geq |a - b|$, we obtain

$$|x - 3| + |x + 1| \geq \left| x - 3 - (x + 1) \right| = 4.$$

Equality occurs when $(x - 3)(x + 1) \leq 0$; that is, when $x \in [-1, 3]$.

Problem 1.40 Show that the equation $|2x - 3| + |x + 1| + |5 - x| = 0.99$ has no solutions.

Solution We use the inequality

$$|\pm a_1 \pm a_2 \pm \cdots \pm a_n| \leq |a_1| + |a_2| + \cdots + |a_n|,$$

which is valid for all numbers $a_1, a_2, \ldots, a_n$. This yields

$$0.99 = |2x - 3| + |x + 1| + |5 - x| \geq \left| (2x - 3) - (x + 1) + (5 - x) \right| = 1,$$

which is impossible. It follows that the given equation has no solutions.

Problem 1.41 Let $a, b > 0$. Find the values of m for which the equation

$$|x - a| + |x - b| + |x + a| + |x + b| = m(a + b)$$

has at least one real solution.

Solution Suppose the equation has at least one real solution x. Then

$$
\begin{aligned}
m(a + b) &= |x - a| + |x + b| + |x - b| + |x + a| \\
&\geq \left| (x - a) - (x + b) \right| + \left| (x - b) - (x + a) \right| = 2(a + b),
\end{aligned}
$$

and since $a + b > 0$, it follows that $m \geq 2$. Conversely, if we assume that $m \geq 2$, then the equation has at least one real solution. Indeed, define

$$f(x) = |x - a| + |x - b| + |x + a| + |x + b|.$$

We observe that $f(0) = 2(a+b) \le m(a+b)$, and $f(ma+mb) = 4m(a+b) > m(a+b)$. The existence of a value x such that $f(x) = m(a+b)$ follows from the fact that f is continuous.

Problem 1.42 Find all possible values of the expression

$$E(x, y, z) = \frac{|x+y|}{|x|+|y|} + \frac{|y+z|}{|y|+|z|} + \frac{|z+x|}{|z|+|x|},$$

where x, y, z are nonzero real numbers.

Solution Two of the three numbers, say x and y, must have the same sign; therefore $|x+y| = |x| + |y|$. It follows that

$$E(x, y, z) = 1 + \frac{|y+z|}{|y|+|z|} + \frac{|z+x|}{|z|+|x|} \ge 1.$$

On the other hand,

$$E(x, y, z) = \frac{|x+y|}{|x|+|y|} + \frac{|y+z|}{|y|+|z|} + \frac{|z+x|}{|z|+|x|} \le 1+1+1 = 3.$$

We show that the set of values of $E(x, y, z)$ is the interval $[1, 3]$. Indeed, $E(1, 1, -1) = 1$ and $E(1, 1, 1) = 3$, and for each $m \in (1, 3)$, we take $x = 1$, $y = z = \frac{m-3}{m+1}$. A short computation shows that in this case, $E(x, y, z) = m$.

Problem 1.43 Find all positive real numbers $x, x_1, x_2, \ldots, x_n$ such that

$$\left| \log(xx_1) \right| + \left| \log(xx_2) \right| + \cdots + \left| \log(xx_n) \right|$$
$$+ \left| \log\left(\frac{x}{x_1}\right) \right| + \left| \log\left(\frac{x}{x_2}\right) \right| + \cdots + \left| \log\left(\frac{x}{x_n}\right) \right|$$
$$= |\log x_1 + \log x_2 + \cdots + \log x_n|.$$

Solution Observe that

$$\left| \log(xx_1) \right| + \left| \log\left(\frac{x}{x_1}\right) \right| \ge \left| \log(xx_1) - \log\left(\frac{x}{x_1}\right) \right| = \left| \log x_1^2 \right| = 2|\log x_1|.$$

It follows that the left-hand side of the equality is greater than or equal to

$$2\big(|\log x_1| + |\log x_2| + \cdots + |\log x_n| \big).$$

However, the right-hand side is less than or equal to

$$|\log x_1| + |\log x_2| + \cdots + |\log x_n|.$$

Thus, equality cannot hold unless

$$|\log x_1| = |\log x_2| = \cdots = |\log x_n| = 0,$$

hence $x_1 = x_2 = \cdots = x_n = 1$. It is not difficult to see that $x = 1$ as well.

Problem 1.44 Prove that for all real numbers a, b, we have

$$\frac{|a+b|}{1+|a+b|} \le \frac{|a|}{1+|a|} + \frac{|b|}{1+|b|}.$$

Solution If a and b have the same sign, then $|a+b| = |a| + |b|$, and we obtain

$$\frac{|a+b|}{1+|a+b|} = \frac{|a|+|b|}{1+|a|+|b|} = \frac{|a|}{1+|a|+|b|} + \frac{|b|}{1+|a|+|b|}$$

$$\le \frac{|a|}{1+|a|} + \frac{|b|}{1+|b|}.$$

Suppose a and b have opposite signs; assume $|a| \ge |b|$. Then $|a+b| \le |a|$, and

$$\frac{|a+b|}{1+|a+b|} \le \frac{|a|}{1+|a|} \le \frac{|a|}{1+|a|} + \frac{|b|}{1+|b|}.$$

Alternatively, we might notice that the function $f(x) = \frac{x}{1+x} = 1 - \frac{1}{1+x}$ is increasing for $x > 0$, and since $|a+b| \le |a| + |b|$, we have

$$\frac{|a+b|}{1+|a+b|} \le \frac{|a|+|b|}{1+|a|+|b|} = \frac{|a|}{1+|a|+|b|} + \frac{|b|}{1+|a|+|b|}$$

$$\le \frac{|a|}{1+|a|} + \frac{|b|}{1+|b|}.$$

Problem 1.45 Let n be an odd positive integer and let $x_1, x_2, \ldots, x_n$ be distinct real numbers. Find all one-to-one functions

$$f : \{x_1, x_2, \ldots, x_n\} \to \{x_1, x_2, \ldots, x_n\}$$

such that

$$|f(x_1) - x_1| = |f(x_2) - x_2| = \cdots = |f(x_n) - x_n|.$$

Solution Let f be a function with the stated property and let

$$a = |f(x_1) - x_1| = |f(x_2) - x_2| = \cdots = |f(x_n) - x_n|.$$

Then for each k, $1 \le k \le n$, we have

$$f(x_k) = x_k + \varepsilon_k a,$$

with either $\varepsilon_k = 1$ or $\varepsilon_k = -1$. Adding these equalities yields

$$\sum_{k=1}^{n} f(x_k) = \sum_{k=1}^{n} x_k + a \sum_{k=1}^{n} \varepsilon_k.$$

Because f is a one-to-one function, $\sum_{k=1}^{n} f(x_k) = \sum_{k=1}^{n} x_k$, and, consequently, $a \sum_{k=1}^{n} \varepsilon_k = 0$. But the sum of an odd number of odd integers cannot be equal to zero, so $\sum_{k=1}^{n} \varepsilon_k \neq 0$. It follows that $a = 0$ and $f(x_k) = x_k$ for all k.

Problem 1.46 Suppose the sequence $a_1, a_2, \ldots, a_n$ satisfies the following conditions:

$$a_1 = 0, \qquad |a_2| = |a_1 + 1|, \ldots, \qquad |a_n| = |a_{n-1} + 1|.$$

Prove that

$$\frac{a_1 + a_2 + \cdots + a_n}{n} \geq -\frac{1}{2}.$$

Solution Define a_{n+1} as follows: $a_{n+1} = |a_n + 1|$. The statements $|x| = |y|$ and $x^2 = y^2$ are equivalent; hence

$$a_2^2 = a_1^2 + 2a_1 + 1,$$
$$a_3^2 = a_2^2 + 2a_2 + 1,$$

$$\vdots$$

$$a_n^2 = a_{n-1}^2 + 2a_{n-1} + 1,$$
$$a_{n+1}^2 = a_n^2 + 2a_n + 1.$$

Adding all these equalities gives

$$a_{n+1}^2 = 2(a_1 + a_2 + \cdots + a_n) + n,$$

so that

$$2(a_1 + a_2 + \cdots + a_n) + n \geq 0,$$

and the result follows.

Problem 1.47 Find real numbers a, b, c such that

$$|ax + by + cz| + |bx + cy + az| + |cx + ay + bz| = |x| + |y| + |z|,$$

for all real numbers x, y, z.

Solution Plugging $x = y = z = 1$ in the given equality yields

$$|a + b + c| = 1.$$

For $x = 1$ and $y = z = 0$, we obtain

$$|a| + |b| + |c| = 1.$$

Since $|a + b + c| = |a| + |b| + |c|$, we deduce that a, b and c have the same sign.

Now, let $x = 1$, $y = -1$ and $c = 0$. It follows that

$$|a - b| + |b - c| + |c - a| = 2.$$

But $|a - b| \leq |a| + |b|$, with equality if and only if a and b have opposite signs or if one of them is zero. Writing the analogous inequalities, we obtain

$$2 = |a - b| + |b - c| + |c - a| \leq 2(|a| + |b| + |c|) = 2.$$

Thus in each pair (a, b), (b, c), (c, a), one of the numbers must be zero, and we conclude that the required triples (a, b, c) are $(1, 0, 0)$, $(0, 1, 0)$, $(0, 0, 1)$, $(-1, 0, 0)$, $(0, -1, 0)$, $(0, 0, -1)$.

4.4 Parameters

Problem 1.50 Solve the equation

$$x = \sqrt{a - \sqrt{a + x}},$$

where $a > 0$ is a parameter.

Solution We must have $x \geq 0$, $a + x \geq 0$ and $a - \sqrt{a + x} \geq 0$. Squaring both sides and rearranging yields

$$x^2 = a - \sqrt{a + x},$$

or

$$\sqrt{a + x} = a - x^2.$$

It follows that $a - x^2 \geq 0$, and thus

$$a + x = (a - x^2)^2,$$

which is equivalent to the fourth degree equation

$$P(x) = 0,$$

where

$$P(x) = x^4 - 2ax^2 - x + a^2 - a.$$

Let us consider a as the unknown and x as the parameter. Then we obtain a quadratic equation in a as follows:

$$a^2 - (2x^2 + 1)a + x^4 - x = 0.$$

Its discriminant is $\Delta = (2x^2 + 1)^2 - 4(x^4 - x) = (2x + 1)^2$; hence the roots are $a_1 = x^2 - x$ and $a_2 = x^2 + x + 1$. This leads to the factorization of P as

$$P(x) = (x^2 - x - a)(x^2 + x + 1 - a).$$

The positive root of the equation $x^2 - x - a = 0$ does not satisfy the condition $a - x^2 \geq 0$, since $a - x^2 = -x$. The roots of the equation $x^2 + x + 1 - a = 0$ are $x_1 = (-1 + \sqrt{4a - 3})/2$ and $x_2 = (-1 - \sqrt{4a - 3})/2$. Only x_1 can be nonnegative, and this happens if and only if $a \geq 1$.

Problem 1.51 Let a be a nonzero real number. Solve the equation

$$a^3 x^4 + 2a^2 x^2 + x + a + 1 = 0.$$

Solution Multiplying the equation by a and setting $t = ax$ yields

$$t^4 + 2at^2 + t + a^2 + a = 0.$$

Considering a as the variable, we obtain the quadratic

$$a^2 + (2t^2 + 1)a + t^4 + t = 0.$$

This is very similar to the equation in the previous problem. We obtain the equations $t^2 + t + a = 0$ and $t^2 - t + 1 + a = 0$. If $a \leq \frac{1}{4}$, then the first of these two equations has solutions $x = \frac{t}{a} = \frac{1}{2a}(-1 \pm \sqrt{(1 - 4a)})$, and if $a \leq -\frac{3}{4}$, then the second equation has solutions $x = \frac{1}{a}(\frac{1}{2} \pm \frac{1}{2}\sqrt{(-3 - 4a)})$.

Problem 1.52 Let $a \in (0, \frac{1}{4})$. Solve the equation

$$x^2 + 2ax + \frac{1}{16} = -a + \sqrt{a^2 + x - \frac{1}{16}}.$$

Solution We write the equation as

$$x^2 + 2ax + a + \frac{1}{16} = \sqrt{a^2 + x - \frac{1}{16}},$$

and square both sides. Rearranging the terms yields

$$x^4 + 4ax^3 + \left(4a^2 + 2a + \frac{1}{8}\right)x^2 + \left(4a^2 + \frac{a}{4} - 1\right)x + \frac{a}{8} + \frac{17}{256} = 0.$$

As before, considering x as a parameter and a as an unknown, we obtain

$$4a^2 x(x + 1) + \frac{1}{8}a(2x + 1)(16x^2 + 1) + x^4 + \frac{1}{8}x^2 - x + \frac{17}{256} = 0.$$

The discriminant of the equation equals

$$\left(\frac{16x^2 + 32x - 1}{8}\right)^2,$$

and the procedure is identical to that of the previous problem.

Observation Another approach is possible. Suppose we view a as the fixed parameter and x as the variable. If we consider the function $f : [-a, +\infty) \to [-a^2 + \frac{1}{16}, +\infty)$, $f(x) = x^2 + 2ax + \frac{1}{16}$, it is not difficult to see that the inverse of f is given by $f^{-1}(x) = -a + \sqrt{a^2 + x - \frac{1}{16}}$; hence the equation

$$f(x) = f^{-1}(x),$$

is equivalent to

$$f(f(x)) = x.$$

We claim that if x_0 is a solution of this equation, then $f(x_0) = x_0$. Indeed, since f is increasing, if $f(x_0) < x_0$, then $x_0 = f(f(x_0)) < f(x_0) < x_0$, and if $f(x_0) > x_0$, then $x_0 = f(f(x_0)) > f(x_0) > x_0$. In both cases we reach a contradiction. The conclusion is that the initial equation is equivalent to

$$x^2 + 2ax + \frac{1}{16} = x,$$

which is a straightforward quadratic equation.

Problem 1.53 Find the positive solutions of the following system of equations:

$$\begin{cases} \frac{a^2}{x^2} - \frac{b^2}{y^2} = 8(y^4 - x^4), \\ ax - by = x^4 - y^4 \end{cases}$$

where $a, b > 0$ are parameters.

Solution We solve the system for a, b instead of x, y. Multiplying the first equation by x^4 yields

$$a^2 x^2 = \frac{b^2 x^4}{y^2} + 8x^4(y^4 - x^4),$$

and from the second equation we deduce

$$a^2 x^2 = (by + x^4 - y^4)^2 = b^2 y^2 + 2by(x^4 - y^4) + (x^4 - y^4)^2.$$

It follows that

$$\frac{b^2(x^4 - y^4)}{y^2} - 2by(x^4 - y^4) - (x^4 - y^4)(9x^4 - y^4) = 0.$$

If $x = y$, from the original system we obtain $a = b$.

If $x \neq y$, we can divide by $x^4 - y^4$ to obtain a quadratic equation in b:

$$b^2 - 2by^3 - y^2(9x^4 - y^4) = 0.$$

The solutions are $y^3 + 3x^2y$ and $y^3 - 3x^2y$. If $b = y^3 - 3x^2y$, the second equation gives $a = x^3 - 3xy^2$. Because $a, b > 0$, it follows that $x^2 > 3y^2$ and $y^2 > 3x^2 > 9y^2$; this is a contradiction. Thus $b = y^3 + 3x^2y$ and $a = x^3 + 3xy^2$. We observe that $a + b = (x + y)^3$ and $a - b = (x - y)^3$; hence $x + y = \sqrt[3]{a + b}$ and $x - y = \sqrt[3]{a - b}$, yielding

$$(x, y) = \left(\frac{\sqrt[3]{a + b} + \sqrt[3]{a - b}}{2}, \frac{\sqrt[3]{a + b} - \sqrt[3]{a - b}}{2} \right).$$

Problem 1.54 Let $a, b, c > 0$. Solve the system of equations

$$\begin{cases} ax - by + \frac{1}{xy} = c, \\ bz - cx + \frac{1}{zx} = a, \\ cy - az + \frac{1}{yz} = b. \end{cases}$$

Solution Considering a, b, c as unknowns, we have a linear system of equations. Multiply the first equation successively by x and y and substitute cx in the second equation and cy in the last one. We obtain

$$\begin{cases} a(x^2 + 1) - b(xy + z) = \frac{1}{xz} - \frac{1}{y}, & \text{(i)} \\ a(xy - z) - b(y^2 + 1) = -\frac{1}{x} - \frac{1}{yz}. & \text{(ii)} \end{cases}$$

Now multiply equation (i) by $y^2 + 1$, and equation (ii) by $-(xy + z)$, and add them up. It follows that

$$a(x^2 + y^2 + z^2 + 1) = \frac{x^2 + y^2 + z^2 + 1}{xz},$$

yielding $a = \frac{1}{xz}$. Analogously, we obtain $b = \frac{1}{yz}$ and $c = \frac{1}{xy}$. Hence $abc = \frac{1}{(xyz)^2}$, so $xyz = \pm \frac{1}{\sqrt{abc}}$. The solutions of the system are

$$\left(\frac{b}{\sqrt{abc}}, \frac{a}{\sqrt{abc}}, \frac{c}{\sqrt{abc}} \right), \left(\frac{-b}{\sqrt{abc}}, \frac{-a}{\sqrt{abc}}, \frac{-c}{\sqrt{abc}} \right).$$

Problem 1.55 Solve the equation

$$x + a^3 = \sqrt[3]{a - x},$$

where a is a real parameter.

Solution Write the equation in the equivalent form

$$\sqrt[3]{\sqrt[3]{a-x}-x}=a.$$

If we consider a as a variable and define the function $f(a) = \sqrt[3]{a-x}$, we see that the equation can be written as

$$f\big(f(a)\big)=a,$$

and since f is an increasing function, this is equivalent to

$$f(a)=a$$

(see the observation in the solution of Problem 1.52). It is not difficult now to obtain the solution $x = a - a^3$.

4.5 Take the Conjugate!

Problem 1.60 Let a and b be distinct positive numbers and let $A = \frac{a+b}{2}$, $B = \sqrt{ab}$. Prove the inequality

$$B < \frac{(a-b)^2}{8(A-B)} < A.$$

Solution Observe that

$$\frac{(a-b)^2}{8(A-B)} = \frac{(a-b)^2(A+B)}{8(A^2-B^2)}$$

and

$$A^2 - B^2 = \frac{a^2+b^2+2ab}{4} - ab = \frac{a^2+b^2-2ab}{4} = \frac{(a-b)^2}{4}.$$

It follows that

$$\frac{(a-b)^2}{8(A-B)} = \frac{A+B}{2},$$

and since $A > B$, we have

$$B < \frac{A+B}{2} < A,$$

as desired.

Problem 1.61 Let m, n be positive integers with $m < n$. Find a closed form for the sum

$$\frac{1}{\sqrt{m}+\sqrt{m+1}} + \frac{1}{\sqrt{m+1}+\sqrt{m+2}} + \cdots + \frac{1}{\sqrt{n-1}+\sqrt{n}}.$$

Solution By taking the conjugate for each term of the sum, we obtain

$$\frac{\sqrt{m+1}-\sqrt{m}}{m+1-m}+\frac{\sqrt{m+2}-\sqrt{m+1}}{m+2-m-1}+\cdots+\frac{\sqrt{n}-\sqrt{n-1}}{n-n+1}$$

$$=\sqrt{m+1}-\sqrt{m}+\sqrt{m+2}-\sqrt{m+1}+\cdots+\sqrt{n}-\sqrt{n-1}=\sqrt{n}-\sqrt{m}.$$

Problem 1.62 For any positive integer n, let

$$f(n)=\frac{4n+\sqrt{4n^2-1}}{\sqrt{2n+1}+\sqrt{2n-1}}.$$

Evaluate the sum $f(1)+f(2)+\cdots+f(40)$.

Solution Let $x=\sqrt{2n+1}$ and $y=\sqrt{2n-1}$. Then $x^2+y^2=4n$, $xy=\sqrt{4n^2-1}$ and $x^2-y^2=2$. Hence

$$f(n)=\frac{x^2+y^2+xy}{x+y}=\frac{x^3-y^3}{x^2-y^2}=\frac{1}{2}\left(\sqrt{(2n+1)^3}-\sqrt{(2n-1)^3}\right).$$

Adding up the equalities for $n=1,2,\ldots,40$ yields

$$f(1)+f(2)+\cdots+f(40)=\frac{1}{2}\left(\sqrt{81^3}-\sqrt{1^3}\right)=364.$$

Problem 1.63 Let a and b be distinct real numbers. Solve the equation

$$\sqrt{x-b^2}-\sqrt{x-a^2}=a-b.$$

Solution We must have $x\geq a^2$ and $x\geq b^2$. Squaring both sides leads to rather complicated computations. We take the conjugate instead, which gives

$$\frac{a^2-b^2}{\sqrt{x-b^2}+\sqrt{x-a^2}}=a-b$$

and we obtain the equivalent form

$$\sqrt{x-b^2}+\sqrt{x-a^2}=a+b.$$

Adding this to the original equation yields

$$\sqrt{x-b^2}=a,$$

and the solution is

$$x=\sqrt{a^2+b^2}.$$

The conditions $x\geq a^2$ and $x\geq b^2$ are clearly satisfied.

Problem 1.64 Solve the following equation, where m is a real parameter:

$$\sqrt{x + \sqrt{x}} - \sqrt{x - \sqrt{x}} = m\sqrt{\frac{x}{x + \sqrt{x}}}.$$

Solution We first notice that x must be a positive number. By taking the conjugate on the left-hand side, we obtain

$$\frac{2\sqrt{x}}{\sqrt{x + \sqrt{x}} + \sqrt{x - \sqrt{x}}} = \frac{m\sqrt{x}}{\sqrt{x + \sqrt{x}}},$$

so m must be a positive number, as well. Simplifying by $\sqrt{x}$ yields the equivalent equation

$$(2 - m)\sqrt{x + \sqrt{x}} = m\sqrt{x - \sqrt{x}}.$$

It follows that $2 - m \geq 0$, and then

$$(2 - m)^2 \left(x + \sqrt{x}\right) = m^2 \left(x - \sqrt{x}\right)$$

or

$$(2 - m)^2 \left(\sqrt{x} + 1\right) = m^2 \left(\sqrt{x} - 1\right).$$

We obtain

$$\sqrt{x} = \frac{m^2 - 2m + 2}{2(m - 1)};$$

therefore $m > 1$ and the solution of the equation is

$$x = \frac{(m^2 - 2m + 2)^2}{4(m - 1)^2},$$

for $1 < m \leq 2$.

Problem 1.65 Prove that for every positive integer k, there exists a positive integer n_k such that $(\sqrt{3} - \sqrt{2})^k = \sqrt{n_k} - \sqrt{n_k - 1}$.

Solution If we take the conjugate in the above equality, we obtain

$$\frac{1}{(\sqrt{3} + \sqrt{2})^k} = \frac{1}{\sqrt{n_k} + \sqrt{n_k - 1}},$$

and hence

$$\left(\sqrt{3} + \sqrt{2}\right)^k = \sqrt{n_k} + \sqrt{n_k - 1}.$$

We deduce that

$$\sqrt{n_k} = \frac{1}{2}\left((\sqrt{3} - \sqrt{2})^k + (\sqrt{3} + \sqrt{2})^k\right)$$

and thus

$$n_k = \frac{1}{4}\left((5 - 2\sqrt{6})^k + (5 + 2\sqrt{6})^k + 2\right).$$

All we have to do now is to prove that the number in the right-hand side is an integer.
Setting $x_k = (5 - 2\sqrt{6})^k + (5 + 2\sqrt{6})^k$, we observe that the sequence $(x_k)_{k \geq 1}$
satisfies the recursive relation:

$$x_{k+2} = 10x_{k+1} - x_k$$

for all $k \geq 0$ (more generally, if $x_k = a^k + b^k$, then $x_{k+2} = (a + b)x_{k+1} - abx_k$,
for all k). Since $x_0 = 2$ and $x_1 = 10$, it follows that x_k is an integer for all
$k \geq 0$. Moreover, using the recursive relation, we can prove inductively that $x_k \equiv$
2 (mod 4) for all k. Indeed, this holds for $k = 0$ and $k = 1$. Assuming that
$x_k \equiv 2$ (mod 4) and $x_{k+1} \equiv 2$ (mod 4), it follows that $x_{k+2} \equiv 10 \cdot 2 - 2 \equiv$
2 (mod 4).

We conclude that $n_k = \frac{1}{4}(x_k + 2)$ is an integer as claimed. Finally, it is not difficult to check that n_k indeed satisfies the given equation.

Problem 1.66 Let a and b be nonzero integers with $|a| \leq 100$, $|b| \leq 100$. Prove
that

$$\left|a\sqrt{2} + b\sqrt{3}\right| \geq \frac{1}{350}.$$

Solution We take the conjugate:

$$\left|a\sqrt{2} + b\sqrt{3}\right| = \frac{|2a^2 - 3b^2|}{|a\sqrt{2} - b\sqrt{3}|},$$

and observe that $2a^2 - 3b^2 \neq 0$; if it were 0, $\sqrt{\frac{3}{2}}$ would be a rational number, a con-
tradiction. Because $2a^2 - 3b^2$ is an integer, it follows that $|2a^2 - 3b^2| \geq 1$. On the
other hand, we have

$$\left|a\sqrt{2} - b\sqrt{3}\right| \leq |a|\sqrt{2} + |b|\sqrt{3} \leq 100(\sqrt{2} + \sqrt{3}) < 350.$$

It follows that

$$\left|a\sqrt{2} + b\sqrt{3}\right| = \frac{|2a^2 - 3b^2|}{|a\sqrt{2} - b\sqrt{3}|} \geq \frac{1}{350}.$$

Problem 1.67 Let n be a positive integer. Prove that

$$\left\lfloor \left(\frac{1 + \sqrt{5}}{2}\right)^{4n-2} \right\rfloor - 1$$

is a perfect square.

Solution Let

$$x_n = \left(\frac{1+\sqrt{5}}{2}\right)^{4n-2} + \left(\frac{1-\sqrt{5}}{2}\right)^{4n-2}.$$

Observe that

$$x_n = \frac{4}{6+2\sqrt{5}}\left(\frac{7+3\sqrt{5}}{2}\right)^{n} + \frac{4}{6-2\sqrt{5}}\left(\frac{7-3\sqrt{5}}{2}\right)^{n},$$

and hence x_n satisfies the recursive relation

$$x_{n+1} = 7x_n - x_{n-1},$$

with $x_0 = x_1 = 3$. It follows that x_n is a positive integer for all $n \geq 0$. Because

$$0 < \left(\frac{1-\sqrt{5}}{2}\right)^{4n-2} < 1,$$

we deduce that

$$\left\lfloor \left(\frac{1+\sqrt{5}}{2}\right)^{4n-2}\right\rfloor - 1 = x_n - 2.$$

But

$$x_n - 2 = \left(\left(\frac{1+\sqrt{5}}{2}\right)^{2n-1} + \left(\frac{1-\sqrt{5}}{2}\right)^{2n-1}\right)^2$$

and a similar argument shows that

$$\left(\frac{1+\sqrt{5}}{2}\right)^{2n-1} + \left(\frac{1-\sqrt{5}}{2}\right)^{2n-1}$$

is a positive integer for all $n \geq 0$. The claim is proved.

Problem 1.68 Consider the sequence

$$a_n = \sqrt{1+\left(1+\frac{1}{n}\right)^2} + \sqrt{1+\left(1-\frac{1}{n}\right)^2}, \quad n \geq 1.$$

Prove that

$$\frac{1}{a_1} + \frac{1}{a_2} + \cdots + \frac{1}{a_{20}}$$

is an integer.

Solution Taking the conjugate, we have

$$\frac{1}{a_n} = \frac{n}{4}\left(\sqrt{1 + \left(1 + \frac{1}{n}\right)^2} - \sqrt{1 + \left(1 - \frac{1}{n}\right)^2}\right)$$

$$= \frac{1}{4}\left(\sqrt{2n^2 + 2n + 1} - \sqrt{2n^2 - 2n + 1}\right).$$

Observe that

$$2(n + 1)^2 - 2(n + 1) + 1 = 2n^2 + 2n + 1,$$

hence the sum telescopes and we have

$$\frac{1}{a_1} + \cdots + \frac{1}{a_{20}} = \frac{1}{4}\left(\sqrt{5} - \sqrt{1} + \sqrt{13} - \sqrt{5} + \cdots + \sqrt{841} - \sqrt{761}\right)$$

$$= \frac{1}{4}(29 - 1) = 7.$$

Problem 1.69 Prove that

$$\sum_{n=1}^{9999} \frac{1}{(\sqrt{n} + \sqrt{n+1})(\sqrt[4]{n} + \sqrt[4]{n+1})} = 9.$$

Solution Observe that

$$\frac{1}{(\sqrt{n} + \sqrt{n+1})(\sqrt[4]{n} + \sqrt[4]{n+1})} = \frac{(\sqrt{n+1} - \sqrt{n})(\sqrt[4]{n+1} - \sqrt[4]{n})}{(n+1-n)(\sqrt{n+1} - \sqrt{n})}$$

$$= \sqrt[4]{n+1} - \sqrt[4]{n}.$$

Thus, the sum telescopes:

$$\sum_{n=1}^{9999} \frac{1}{(\sqrt{n} + \sqrt{n+1})(\sqrt[4]{n} + \sqrt[4]{n+1})} = \sum_{n=1}^{9999} \left(\sqrt[4]{n+1} - \sqrt[4]{n}\right)$$

$$= \sqrt[4]{10000} - \sqrt[4]{1}$$

$$= 9.$$

Problem 1.70 Consider the sequence

$$a_n = 2 - \frac{1}{n^2 + \sqrt{n^4 + \frac{1}{4}}}, \quad n \geq 1.$$

Prove that

$$\sqrt{a_1} + \sqrt{a_2} + \cdots + \sqrt{a_{119}}$$

is an integer.

Solution We have

$$a_n = 2 - \frac{n^2 - \sqrt{n^4 + \frac{1}{4}}}{-\frac{1}{4}} = 2 + 4n^2 - 2\sqrt{4n^4 + 1}.$$

But

$$4n^4 + 1 = 4n^4 + 4n^2 + 1 - 4n^2$$

$$= \left(2n^2 + 1\right)^2 - 4n^2$$

$$= \left(2n^2 + 2n + 1\right)\left(2n^2 - 2n + 1\right),$$

and

$$2 + 4n^2 = \left(2n^2 + 2n + 1\right) + \left(2n^2 - 2n + 1\right).$$

Therefore,

$$2 + 4n^2 - 2\sqrt{4n^4 + 1} = \left(\sqrt{2n^2 + 2n + 1} - \sqrt{2n^2 - 2n + 1}\right)^2,$$

and

$$\sqrt{a_n} = \sqrt{2n^2 + 2n + 1} - \sqrt{2n^2 - 2n + 1}.$$

Again, since

$$2(n + 1)^2 - 2(n + 1) + 1 = 2n^2 + 2n + 1,$$

the sum telescopes and we obtain

$$\sqrt{a_1} + \sqrt{a_2} + \cdots + \sqrt{a_{119}} = \sqrt{2 \cdot 119^2 + 2 \cdot 119 + 1} - \sqrt{2 \cdot 1^2 - 2 \cdot 1 + 1}$$

$$= 169 - 1 = 168,$$

an integer number.

4.6 Inequalities with Convex Functions

Problem 1.74 Let $a, b > 0$ and let n be a positive integer. Prove the inequality

$$\frac{a^n + b^n}{2} \geq \left(\frac{a + b}{2}\right)^n.$$

Solution We can try induction on n. The assertion holds for $n = 1$ and, assuming it holds for n, multiply both sides by $\frac{a+b}{2}$. It follows that

$$\frac{(a^n + b^n)(a+b)}{4} \geq \left(\frac{a+b}{2}\right)^{n+1},$$

and therefore it is sufficient to prove that

$$\frac{a^{n+1} + b^{n+1}}{2} \geq \frac{(a^n + b^n)(a+b)}{4}.$$

A short computation shows that this is equivalent to

$$(a - b)(a^n - b^n) \geq 0,$$

which is true since $a - b$ and $a^n - b^n$ have the same sign.

A shorter solution uses convex functions. Consider $n \geq 2$ and let $f : (0, +\infty) \to (0, +\infty)$; $f(x) = x^n$. Because $f''(x) = n(n-1)x^{n-2} > 0$, f is convex; thus for all $a, b > 0$ we have

$$\frac{f(a) + f(b)}{2} \geq f\left(\frac{a+b}{2}\right).$$

This is exactly the inequality we had to prove.

Problem 1.75 Prove that

$$\sqrt[3]{3 - \sqrt[3]{3}} + \sqrt[3]{3 + \sqrt[3]{3}} < 2\sqrt[3]{3}.$$

Solution Consider the function $f(x) = \sqrt[3]{x}$. It is not difficult to see that f is concave on the interval $(0, +\infty)$ and it is not linear. Then, for any two distinct numbers $a, b > 0$ we have

$$\frac{f(a) + f(b)}{2} < f\left(\frac{a+b}{2}\right).$$

If we take $a = 3 - \sqrt[3]{3}$ and $b = 3 + \sqrt[3]{3}$, we obtain the desired result.

Problem 1.76 Prove the $AM - GM$ inequality

$$\frac{x_1 + x_2 + \cdots + x_n}{n} \geq \sqrt[n]{x_1 x_2 \cdots x_n},$$

for all $x_1, x_2, \ldots, x_n > 0$.

Solution There are many proofs of the $AM - GM$ inequality. One of them uses convexity. Consider the function $f : (0, +\infty) \to (0, +\infty)$, $f(x) = \log x$. This is a concave function, so we have the inequality

$$\frac{f(x_1) + f(x_2) + \cdots + f(x_n)}{n} \leq f\left(\frac{x_1 + x_2 + \cdots + x_n}{n}\right),$$

for all $x_1, x_2, \ldots, x_n > 0$. Using the properties of the logarithmic function, we obtain

$$\frac{\log x_1 + \log x_2 + \cdots + \log x_n}{n} = \frac{\log(x_1 x_2 \cdots x_n)}{n}$$

$$= \log(x_1 x_2 \cdots x_n)^{\frac{1}{n}}$$

$$= \log \sqrt[n]{x_1 x_2 \cdots x_n},$$

and hence

$$\log \sqrt[n]{x_1 x_2 \cdots x_n} \le \log \frac{x_1 + x_2 + \cdots + x_n}{n}.$$

Because f is increasing, this is equivalent to the $AM - GM$ inequality.

Problem 1.77 Let $a_1 < a_2 < \cdots < a_{2n+1}$ be positive real numbers. Prove the inequality

$$\sqrt[n]{a_1 - a_2 + a_3 - \cdots - a_{2n} + a_{2n+1}} \ge \sqrt[n]{a_1} - \sqrt[n]{a_2} + \cdots + \sqrt[n]{a_{2n+1}}.$$

Solution We will prove inductively a stronger statement. Let $m \ge 2$ and $n \ge 1$ be integers. Then the following inequality holds for any positive real numbers $a_1 < a_2 < \cdots < a_{2n+1}$:

$$\sqrt[m]{a_1 - a_2 + a_3 - \cdots - a_{2n} + a_{2n+1}} \ge \sqrt[m]{a_1} - \sqrt[m]{a_2} + \cdots + \sqrt[m]{a_{2n+1}}.$$

Now, we fix m and prove the assertion by induction on n. For $n = 1$ we have to prove that

$$\sqrt[m]{a_1 - a_2 + a_3} \ge \sqrt[m]{a_1} - \sqrt[m]{a_2} + \sqrt[m]{a_3},$$

for all $0 < a_1 < a_2 < a_3$. This follows from problem 1, for the concave function $f : (0, +\infty) \to (0, +\infty)$, $f(x) = \sqrt[m]{x}$. Suppose the inequality holds for n and let $0 < a_1 < a_2 < \cdots < a_{2n+1} < a_{2n+2} < a_{2n+3}$. Then, from the induction hypothesis, we have

$$\sqrt[m]{a_1 - a_2 + \cdots + a_{2n+1} - a_{2n+2} + a_{2n+3}}$$

$$\ge \sqrt[m]{a_1} - \sqrt[m]{a_2} + \cdots + \sqrt[m]{a_{2n+1} - a_{2n+2} + a_{2n+3}},$$

since $a_1 < a_2 < \cdots < a_{2n} < a_{2n+1} - a_{2n+2} + a_{2n+3}$. But

$$\sqrt[m]{a_{2n+1} - a_{2n+2} + a_{2n+3}} \ge \sqrt[m]{a_{2n+1}} - \sqrt[m]{a_{2n+2}} + \sqrt[m]{a_{2n+3}},$$

and the result follows.

Problem 1.78 Let $x, y, z > 0$. Prove that

$$\frac{x}{2x + y + z} + \frac{y}{x + 2y + z} + \frac{z}{x + y + 2z} \le \frac{3}{4}.$$

Solution Let $s = x + y + z$. The inequality becomes

$$\frac{x}{s+x} + \frac{y}{s+y} + \frac{z}{s+z} \leq \frac{3}{4}.$$

Consider the function $f : (0, +\infty) \to (0, +\infty)$ defined by

$$f(t) = \frac{t}{s+t}.$$

We see that

$$f''(x) = -\frac{2s}{(s+t)^3} < 0,$$

for all $t > 0$, so f is concave. It follows that

$$\frac{f(x) + f(y) + f(z)}{3} \leq f\left(\frac{x+y+z}{3}\right).$$

But

$$f\left(\frac{x+y+z}{3}\right) = f\left(\frac{s}{3}\right) = \frac{\frac{s}{3}}{s+\frac{s}{3}} = \frac{1}{4},$$

and thus

$$f(x) + f(y) + f(z) \leq \frac{3}{4},$$

as desired.

Problem 1.79 Prove that if $a, b, c, d > 0$ and $a \leq 1, a + b \leq 5, a + b + c \leq 14$, $a + b + c + d \leq 30$, then

$$\sqrt{a} + \sqrt{b} + \sqrt{c} + \sqrt{d} \leq 10.$$

Solution The function $f : (0, +\infty) \to (0, +\infty)$, defined by $f(x) = \sqrt{x}$, is concave, and therefore, for any positive real numbers $\lambda_1, \lambda_2, \ldots, \lambda_n$ such that $\lambda_1 + \lambda_2 + \cdots + \lambda_n = 1$, we have

$$\lambda_1 f(x_1) + \lambda_2 f(x_2) + \cdots + \lambda_n f(x_n) \leq f(\lambda_1 x_1 + \lambda_2 x_2 + \cdots + \lambda_n x_n).$$

Now, take $n = 4$ and $\lambda_1 = \frac{1}{10}, \lambda_2 = \frac{2}{10}, \lambda_3 = \frac{3}{10}, \lambda_4 = \frac{4}{10}$. It follows that

$$\frac{1}{10}\sqrt{a} + \frac{2}{10}\sqrt{\frac{b}{4}} + \frac{3}{10}\sqrt{\frac{c}{9}} + \frac{4}{10}\sqrt{\frac{d}{16}} \leq \sqrt{\frac{a}{10} + \frac{b}{20} + \frac{c}{30} + \frac{d}{40}},$$

so

$$\sqrt{a} + \sqrt{b} + \sqrt{c} + \sqrt{d} \leq 10\sqrt{\frac{12a + 6b + 4c + 3d}{120}}.$$

But

$$12a + 6b + 4c + 3d = 3(a + b + c + d) + (a + b + c) + 2(a + b) + 6a$$
$$\leq 3 \cdot 30 + 14 + 2 \cdot 5 + 6 \cdot 1 = 120,$$

and the claim is proved.

4.7 Induction at Work

Problem 1.83 Let n be a positive integer. Prove the inequality

$$\left(1 + \frac{1}{2}\right)\left(1 + \frac{1}{2^2}\right)\left(1 + \frac{1}{2^3}\right) \cdots \left(1 + \frac{1}{2^n}\right) < \frac{5}{2}.$$

Solution As in Problem 1.81, we will prove a stronger statement:

$$\left(1 + \frac{1}{2}\right)\left(1 + \frac{1}{2^2}\right)\left(1 + \frac{1}{2^3}\right) \cdots \left(1 + \frac{1}{2^n}\right) \leq \frac{5}{2}\left(1 - \frac{1}{2^n}\right)$$

for all $n \geq 2$. For $n = 2$, both sides of the above inequality equal $\frac{15}{8}$. Suppose the inequality holds for $n > 2$ and multiply both sides by $1 + \frac{1}{2^{n+1}}$. We have to prove that

$$\frac{5}{2}\left(1 - \frac{1}{2^n}\right)\left(1 + \frac{1}{2^{n+1}}\right) \leq \frac{5}{2}\left(1 - \frac{1}{2^{n+1}}\right),$$

and a short computation shows that the latter is equivalent to the obvious $\frac{1}{2^{2n+1}} > 0$.

Problem 1.84 Let n be a positive integer. Prove that the number

$$a_n = 2^{2^n} - 1$$

has at least n distinct prime divisors.

Solution Note that $a_1 = 3$ has one prime divisor and suppose that for some $n > 1$, a_n has at least n distinct prime divisors $p_1, p_2, \ldots, p_n$. Now,

$$a_{n+1} = 2^{2^{n+1}} - 1 = \left(2^{2^n}\right)^2 - 1 = a_n\left(2^{2^n} + 1\right),$$

and hence $p_1, p_2, \ldots, p_n$ are distinct prime divisors of a_{n+1} as well. The numbers $2^{2^n} + 1$ and $2^{2^n} - 1$ are coprime since any of their common divisors greater than 1 must be odd and must divide $2^{2^n} + 1 - (2^{2^n} - 1) = 2$, which is impossible. Therefore any prime divisor of $2^{2^n} + 1$ is different from $p_1, p_2, \ldots, p_n$. It follows that a_{n+1} has at least $n + 1$ distinct prime divisors.

Problem 1.85 Let a and n be positive integers such that a is less than or equal to $n!$. Prove that a can be represented as a sum of at most n distinct divisors of $n!$.

Solution The assertion is trivial for $n = 1$, so suppose it holds true for some $n > 1$. We have $a \le (n + 1)!$. By the division algorithm, there exist b, c such that $a = b(n + 1) + c$, with $0 \le c \le n$. Clearly, $b \le n!$; and thus, using the induction hypothesis, b can be represented as a sum of at most n distinct divisors of $n!$, say

$$b = d_1 + d_2 + \cdots + d_k,$$

with $k \le n$. But then

$$a = d_1(n + 1) + d_2(n + 1) + \cdots + d_k(n + 1) + c,$$

and the numbers $d_1(n + 1), d_2(n + 1), \ldots, d_k(n + 1)$ and c are distinct divisors of $(n + 1)!$.

Problem 1.86 Let $x_1, x_2, \ldots, x_m, y_1, y_2, \ldots, y_n$ be positive integers such that the sums $x_1 + x_2 + \cdots + x_m$ and $y_1 + y_2 + \cdots + y_n$ are equal and less than mn. Prove that in the equality

$$x_1 + x_2 + \cdots + x_m = y_1 + y_2 + \cdots + y_n$$

one can cancel some terms and obtain another equality.

Solution Let $k = m + n$. We proceed by induction on k. Let $s = x_1 + x_2 + \cdots + x_m = y_1 + y_2 + \cdots + y_n$. Since $s \ge m$, $s \ge n$ and $s < mn$, it follows that $m, n \ge 2$, and hence $k \ge 4$. The case $k = 4$ is easy to discard, so let us suppose that $k > 4$. With no loss of generality, we may assume that $x_1 \ge x_2 \ge \cdots \ge x_m$ and $y_1 \ge y_2 \ge \cdots \ge y_n$. If $x_1 = y_1$, these terms cancel and we are done. Suppose, for instance, that $x_1 > y_1$. We want to apply the induction hypothesis to the equality

$$(x_1 - y_1) + x_2 + \cdots + x_m = y_2 + y_3 + \cdots + y_n$$

in which there are $m + (n - 1) = k - 1$ terms. For this, we have to check that the sum $s' = y_2 + y_3 + \cdots + y_n$ verifies the condition $s' < m(n - 1)$. Because $y_1 \ge y_2 \ge \cdots \ge y_n$, it follows that $y_1 \ge \frac{s}{n}$, so that

$$s' = s - y_1 \le s - \frac{s}{n} = s\frac{(n - 1)}{n} < mn\frac{(n - 1)}{n} = m(n - 1),$$

as needed.

Problem 1.87 The sequence $(x_n)_{n \ge 1}$ is defined by $x_1 = 1$, $x_{2n} = 1 + x_n$ and $x_{2n+1} = \frac{1}{x_{2n}}$ for all $n \ge 1$. Prove that for any positive rational number r there exists an unique n such that $r = x_n$.

Solution Note that all the terms of the sequence are positive numbers and that $x_{2n} > 1$, $x_{2n+1} < 1$ for all $n \geq 1$. We prove by induction on $k \geq 2$ the following statement: for all positive integers a, b such that $\gcd(a, b) = 1$ and $a + b \leq k$, there exists a term of the sequence equal to $\frac{a}{b}$. If $k = 2$, then $a = b = 1$ and $\frac{a}{b} = 1 = x_1$. Suppose that the statement is true for some $k > 2$, and let a, b be coprime positive integers such that $a + b = k + 1$. If $a > b$, then we apply the induction hypothesis to the numbers $a - b$ and b. Clearly, $\gcd(a - b, b) = 1$ and $(a - b) + b = a \leq k$; therefore, there exists n such that $x_n = \frac{a-b}{b}$. But then

$$x_{2n} = 1 + x_n = 1 + \frac{a - b}{b} = \frac{a}{b}.$$

If $a < b$, then we apply the induction hypothesis to the numbers $b - a$ and a. There exists n such that $x_n = \frac{b-a}{a}$, and we obtain

$$x_{2n+1} = \frac{1}{x_{2n}} = \frac{1}{1 + x_n} = \frac{1}{1 + \frac{b-a}{a}} = \frac{a}{b}.$$

We proved that for any positive rational number r there exists n such that $r = x_n$, but we still have to prove that n is unique. This follows from the fact that the terms of the sequence $(x_n)_{n \geq 1}$ are pairwise distinct. Indeed, suppose not, and choose $n \neq m$ such that $x_n = x_m$, with minimal n. The observation at the beginning of the solution shows that n and m have the same parity. If $n = 2n'$ and $m = 2m'$, then we obtain $x_{n'} = x_{m'}$, contradicting the minimality of n. The case $n = 2n' + 1$ and $m = 2m' + 1$ yields $x_{2n'} = x_{2m'}$, and we again reach a contradiction.

Problem 1.88 Let n be a positive integer and let $0 < a_1 < a_2 < \cdots < a_n$ be real numbers. Prove that at least $\binom{n+1}{2}$ of the sums $\pm a_1 \pm a_2 \pm \cdots \pm a_n$ are distinct.

Solution The base case is obvious since for $n = 1$ we have two sums: a_1 and $-a_1$. Suppose the assertion true for some k and consider the numbers $0 < a_1 < a_2 < \cdots < a_k < a_{k+1}$. The greatest sum determined by the first k numbers is $s = a_1 + a_2 + \cdots + a_k$. Consider the sums

$$s + a_{k+1} > s + a_{k+1} - a_1 > s + a_{k+1} - a_2 > \cdots > s + a_{k+1} - a_k.$$

We have $s + a_{k+1} - a_i > s$, for all i, hence all these $k + 1$ sums are distinct from the sums determined by the first k numbers. The proof ends by noticing that

$$\binom{k+1}{2} + k + 1 = \binom{k+2}{2}.$$

Problem 1.89 Prove that for each positive integer n, there are pairwise relatively prime integers $k_0, k_1, \ldots, k_n$, all strictly greater than 1, such that $k_0 k_1 \cdots k_n - 1$ is the product of two consecutive integers.

Solution For $n = 1$ we can choose $k_0 = 3$ and $k_1 = 7$ since $3 \cdot 7 - 1 = 4 \cdot 5$. Suppose that for $n > 1$ there exist pairwise relatively prime integers $1 < k_0 < k_1 < \cdots < k_n$ such that

$$k_0 k_1 \cdots k_n - 1 = a_n(a_n - 1),$$

for some integer a_n. If we choose $k_{n+1} = a_n^2 + a_n + 1$, then

$$k_0 k_1 \cdots k_n k_{n+1} = (a_n^2 - a_n + 1)(a_n^2 + a_n + 1) = a_n^4 + a_n^2 + 1,$$

and hence

$$k_0 k_1 \cdots k_n k_{n+1} - 1 = a_n^2(a_n^2 + 1)$$

is indeed the product of two consecutive integers.

Now, suppose that $\gcd(k_{n+1}, k_j) = d$, for some j, $1 \le j \le n$. Clearly, d must be an odd integer, since k_{n+1} is odd. Then $d \mid k_1 k_2 \cdots k_n = a_n^2 - a_n + 1$ and $d \mid a_n^2 + a_n + 1$ hence $d \mid 2a_n$ and since d is odd, $d \mid a_n$. From $d \mid a_n$ and $d \mid a_n^2 + a_n + 1$ we deduce that $d = 1$, therefore $k_1, k_2, \ldots, k_n, k_{n+1}$ are pairwise relatively prime integers.

Problem 1.90 Prove that for every positive integer n, the number $3^{3^n} + 1$ is the product of at least $2n + 1$ (not necessarily distinct) primes.

Solution For $n = 1$ we have $3^{3^1} + 1 = 28 = 2 \cdot 2 \cdot 7$, as needed. Suppose that $3^{3^n} + 1$ is the product of at least $2n + 1$ (not necessarily distinct) primes and observe that

$$3^{3^{n+1}} + 1 = (3^{3^n})^3 + 1 = (3^{3^n} + 1)((3^{3^n})^2 - 3^{3^n} + 1),$$

hence it suffices to show that $(3^{3^n})^2 - 3^{3^n} + 1$ is the product of at least two primes.
Using the obvious identity $x^2 - x + 1 = (x + 1)^2 - 3x$, we have

$$\begin{aligned}
(3^{3^n})^2 - 3^{3^n} + 1 &= (3^{3^n} + 1)^2 - 3 \cdot 3^{3^n} \\
&= (3^{3^n} + 1)^2 - 3^{3^n+1} \\
&= (3^{3^n} + 1)^2 - (3^{\frac{3^n+1}{2}})^2 \\
&= (3^{3^n} + 1 - 3^{\frac{3^n+1}{2}})(3^{3^n} + 1 + 3^{\frac{3^n+1}{2}}).
\end{aligned}$$

Since $3^{3^n} + 1 - 3^{\frac{3^n+1}{2}} > 1$, the conclusion follows.

Problem 1.91 Prove that for every positive integer n there exists an n-digit number divisible by 5^n all of whose digits are odd.

Solution Clearly, the assertion holds for $n = 1$. Assume now that the number $A = a_1 a_2 \cdots a_n = 5^n \cdot a$, for some integer a, and $a_1, a_2, \ldots, a_n$ are all odd. Consider the following five numbers

$$A_1 = 1 a_1 a_2 \cdots a_n = 1 \cdot 10^n + 5^n \cdot a = 5^n(1 \cdot 2^n + a),$$

$$A_2 = 3a_1a_2 \cdots a_n = 3 \cdot 10^n + 5^n \cdot a = 5^n \left(3 \cdot 2^n + a\right),$$

$$A_3 = 5a_1a_2 \cdots a_n = 5 \cdot 10^n + 5^n \cdot a = 5^n \left(5 \cdot 2^n + a\right),$$

$$A_4 = 7a_1a_2 \cdots a_n = 7 \cdot 10^n + 5^n \cdot a = 5^n \left(7 \cdot 2^n + a\right),$$

$$A_5 = 9a_1a_2 \cdots a_n = 9 \cdot 10^n + 5^n \cdot a = 5^n \left(9 \cdot 2^n + a\right).$$

If two of the numbers $1 \cdot 2^n + a$, $3 \cdot 2^n + a$, $5 \cdot 2^n + a$, $7 \cdot 2^n + a$, and $9 \cdot 2^n + a$ give the same remainder when divided by 5, then their difference must be divisible by 5, which is clearly impossible. It follows that the five numbers give distinct remainders when divided by 5, hence one of them is divisible by 5. We conclude that one of the numbers A_1, A_2, A_3, A_4, A_5 is divisible by $5 \cdot 5^n = 5^{n+1}$, as desired.

4.8 Roots and Coefficients

Problem 1.95 Let a, b, c be nonzero real numbers such that $a + b + c \neq 0$ and

$$\frac{1}{a} + \frac{1}{b} + \frac{1}{c} = \frac{1}{a+b+c}.$$

Prove that for all odd integers n

$$\frac{1}{a^n} + \frac{1}{b^n} + \frac{1}{c^n} = \frac{1}{a^n + b^n + c^n}.$$

Solution Let $f(X) = X^3 + mX^2 + nX + p$ be the monic polynomial with roots a, b, c. The given equality can be written as

$$(a+b+c)(ab+bc+ca) = abc,$$

so, by Viète's relations, we obtain $p = mn$. Thus

$$f(X) = X^3 + mX^2 + nX + mn = X^2(X+m) + n(X+m)$$
$$= (X+m)\left(X^2+n\right).$$

It follows that one of the roots, say a, equals $-m = a + b + c$; hence $b + c = 0$. Substituting $c = -b$, the desired equality becomes

$$\frac{1}{a^n} + \frac{1}{b^n} + \frac{1}{(-b)^n} = \frac{1}{a^n + b^n + (-b)^n},$$

which is clearly true if n is an odd integer.

Observation We could also ask the following question: if m and n are odd integers, prove that

$$\frac{1}{a^m} + \frac{1}{b^m} + \frac{1}{c^m} = \frac{1}{a^m + b^m + c^m}$$

if and only if

$$\frac{1}{a^n} + \frac{1}{b^n} + \frac{1}{c^n} = \frac{1}{a^n + b^n + c^n}.$$

Problem 1.96 Let $a \leq b \leq c$ be real numbers such that

$$a + b + c = 2$$

and

$$ab + bc + ca = 1.$$

Prove that

$$0 \leq a \leq \frac{1}{3} \leq b \leq 1 \leq c \leq \frac{4}{3}.$$

Solution Let $f(X) = X^3 + mX^2 + nX + p$ be the monic polynomial with roots a, b, c. Viète's relations yield $m = -2$ and $n = 1$, hence

$$f(X) = X^3 - 2X^2 + X + p.$$

The derivative of f is

$$f'(X) = 3X^2 - 4X + 1,$$

with roots $\frac{1}{3}$ and 1. Using the sign of f', we deduce that f increases on the interval $(-\infty, \frac{1}{3}]$ from $-\infty$ to $f(\frac{1}{3}) = p + \frac{4}{27}$, and then decreases on the interval $[\frac{1}{3}, 1]$ from $p + \frac{4}{27}$ to $f(1) = p$. Finally, f increases again on the interval $[1, +\infty)$ from p to $+\infty$. It follows that f has three real roots if and only if $p + \frac{4}{27} \geq 0$ and $p \leq 0$.

Observe that $f(0) = p$ and $f(\frac{4}{3}) = p + \frac{4}{27}$, and hence f changes its sign on each of the intervals $[0, \frac{1}{3}], [\frac{1}{3}, 1]$ and $[1, \frac{4}{3}]$. We deduce that each of these intervals contains a root of f. The claim follows.

Problem 1.97 Prove that two of the four roots of the polynomial $X^4 + 12X - 5$ add up to 2.

Solution Let x_1, x_2, x_3 and x_4 be the roots of the polynomial. Writing Viète's relations in terms of $s = x_1 + x_2$, $s' = x_3 + x_4$, $p = x_1 x_2$ and $p' = x_3 x_4$, we obtain

$$s + s' = 0,$$
$$p + p' + ss' = 0,$$
$$ps' + p's = -12,$$
$$pp' = -5.$$

Substituting $s' = -s$ into the second and third equalities yields

$$p + p' = s^2,$$

$$-p + p' = -\frac{12}{s},$$

and we obtain

$$p = \frac{1}{2}\left(s^2 + \frac{12}{s}\right), \qquad p' = \frac{1}{2}\left(s^2 - \frac{12}{s}\right).$$

From $pp' = -5$ it follows that

$$\frac{1}{4}\left(s^2 + \frac{12}{s}\right)\left(s^2 - \frac{12}{s}\right) = -5,$$

or

$$s^6 + 20s^2 - 144 = 0.$$

Hence $s = x_1 + x_2$ is a root of the polynomial

$$P(X) = X^6 + 20X^2 - 144.$$

Using similar arguments, we deduce that its other five roots are $x_1 + x_3, x_1 + x_4, x_2 + x_3, x_2 + x_4$ and $x_3 + x_4$. It suffices now to check that 2 is a root of P. Indeed,

$$P(2) = 64 + 20 \cdot 4 - 144 = 0.$$

Problem 1.98 Find m and solve the following equation, knowing that its roots form a geometrical sequence

$$X^4 - 15X^3 + 70X^2 - 120X + m = 0.$$

Solution Let x_1, x_2, x_3, x_4 be the roots of the polynomial and suppose they form a geometrical sequence in the order x_1, x_3, x_4, x_2. We again write Viète's relations in terms of $s = x_1 + x_2$, $s' = x_3 + x_4$, $p = x_1 x_2$ and $p' = x_3 x_4$. This gives

$$s + s' = 15,$$
$$p + p' + ss' = 70,$$
$$ps' + p's = 120,$$
$$pp' = m.$$

Then $x_1 x_2 = x_3 x_4$, and hence $p = p'$. It follows that $120 = ps' + p's = p(s + s') = 15p$ and thus $p = 8$. Then $m = 64$, and $ss' = 70 - 16 = 54$. The numbers s and s' are the roots of the equation $X^2 - 15X + 54 = 0$, hence $s = 6$ and $s' = 9$ or vice versa. It follows that the roots of the polynomial are $1, 2, 4$ and 8.

Problem 1.99 Let $x_1, x_2, \ldots, x_n$ be the roots of the polynomial $X^n + X^{n-1} + \cdots + X + 1$. Prove that

$$\frac{1}{1 - x_1} + \frac{1}{1 - x_2} + \cdots + \frac{1}{1 - x_n} = \frac{n}{2}.$$

Solution We look for the polynomial with roots

$$y_k = \frac{1}{1 - x_k}, \quad k = 1, 2, \ldots, n.$$

From the above equality, it follows that

$$x_k = \frac{y_k - 1}{y_k},$$

and since x_k is a root of $X^n + X^{n-1} + \cdots + X + 1$, we obtain

$$\left(\frac{y_k - 1}{y_k}\right)^n + \left(\frac{y_k - 1}{y_k}\right)^{n-1} + \cdots + \frac{y_k - 1}{y_k} + 1 = 0.$$

This is equivalent to

$$(y_k - 1)^n + y_k(y_k - 1)^{n-1} + \cdots + y_k^{n-1}(y_k - 1) + y_k^n = 0.$$

It follows that y_k is a root of the polynomial

$$P(X) = (X - 1)^n + X(X - 1)^{n-1} + \cdots + X^{n-1}(X - 1) + X^n.$$

The desired sum equals

$$y_1 + y_2 + \cdots + y_n,$$

and its value can be determined by using Viète's relations. Observe that

$$P(X) = (n + 1)X^n - X^{n-1}\left(\binom{n}{1} + \binom{n-1}{1} + \cdots + \binom{1}{1}\right) + \cdots$$

and hence

$$y_1 + y_2 + \cdots + y_n = \frac{\binom{n}{1} + \binom{n-1}{1} + \cdots + \binom{1}{1}}{n + 1} = \frac{n(n + 1)}{2(n + 1)} = \frac{n}{2},$$

as claimed.

Problem 1.100 Let a, b, c be rational numbers and let x_1, x_2, x_3 be the roots of the polynomial $P(X) = X^3 + aX^2 + bX + c$. Prove that if $\frac{x_1}{x_2}$ is a rational number, different from 0 and -1, then x_1, x_2, x_3 are rational numbers.

Solution Let $\frac{x_1}{x_2} = m \in \mathbf{Q}, m \neq 0, -1$. We have

$$x_1 + x_2 + x_3 = -a.$$

We claim that if one of the roots is a rational number, then so are all three of them. Indeed, if $x_1 \in \mathbf{Q}$, then $x_2 = \frac{x_1}{m} \in \mathbf{Q}$ and $x_3 = -a - x_1 - x_2 \in \mathbf{Q}$. If $x_2 \in \mathbf{Q}$, then

$x_1 = mx_2 \in \mathbf{Q}$ and $x_3 \in \mathbf{Q}$. If $x_3 \in \mathbf{Q}$, then $x_2 = \frac{-a-x_3}{1+m} \in \mathbf{Q}$ and $x_1 = mx_2 \in \mathbf{Q}$. It suffices now to prove that P has a rational root.

Substituting $x_1 = mx_2$ and $x_3 = -a - mx_2 - x_2$ in the equality

$$x_1x_2 + x_1x_3 + x_2x_3 = b$$

yields

$$\left(m^2 + m + 1\right)x_2^2 + a(m+1)x_2 + b = 0,$$

and consequently x_2 is a root of a second degree polynomial f with rational coefficients. Since P and f share a common root, their greatest common divisor is a nonconstant polynomial with rational coefficients. It follows that P can be decomposed into factors with rational coefficients; since one of the factors must be of degree one, it has at least one rational root.

Problem 1.101 Solve in real numbers the system

$$\begin{cases} x + y + z = 0, \\ x^3 + y^3 + z^3 = 18, \\ x^7 + y^7 + z^7 = 2058. \end{cases}$$

Solution Consider the polynomial

$$P(t) = t^3 + at^2 + bt + c,$$

with roots x, y, z.

Since $x + y + z = 0$, it follows that $a = 0$, hence

$$P(t) = t^3 + bt + c.$$

Because x, y, z are the roots of P, we have

$$x^3 + bx + c = 0,$$
$$y^3 + by + c = 0,$$
$$z^3 + bz + c = 0.$$

Adding these equalities and using the fact that $x^3 + y^3 + z^3 = 18$, we obtain $c = -6$. Therefore

$$P(t) = t^3 + bt - 6.$$

Now, use the last equation of the system to find b. Multiply the previous equalities by x^n, y^n, z^n, respectively, and add them to obtain

$$x^{n+3} + y^{n+3} + z^{n+3} + b\left(x^{n+1} + y^{n+1} + z^{n+1}\right) - 6\left(x^n + y^n + z^n\right) = 0.$$

Denoting $S_n = x^n + y^n + z^n$ for all $n \geq 1$, this equality becomes

$$S_{n+3} + bS_{n+1} - 6S_n = 0 \qquad\qquad (*)$$

for all positive integers n. We have $S_7 = 2058$. On the other hand, using $(*)$, we obtain

$$S_7 = -bS_5 + 6S_4 = -b(-bS_3 + 6S_2) + 6(-bS_2 + 6S_1)$$
$$= b^2 S_3 - 12bS_2 + 36S_1.$$

Since $S_3 = 18$, $S_2 = (x + y + z)^2 - 2(xy + xz + yz) = -2b$ and $S_1 = 0$, it follows that

$$S_7 = 42b^2,$$

so $b = \pm 7$.

The equation $t^3 + 7t - 6 = 0$ has only one real root (observe that the function $f(t) = t^3 + 7t - 6$ is strictly increasing). The equation $t^3 - 7t - 6 = 0$ has roots $t_1 = -1$, $t_2 = -2$ and $t_3 = 3$, and thus the solutions of the system are $(-1, -2, 3)$ and all of its permutations.

Problem 1.102 Solve in real numbers the system of equations

$$\begin{cases} a + b = 8, \\ ab + c + d = 23, \\ ad + bc = 28, \\ cd = 12. \end{cases}$$

Solution The expressions on the left-hand side remind us of the coefficients obtained when two polynomials are multiplied. Indeed, observe that

$$(x^2 + ax + c)(x^2 + bx + d) = x^4 + (a+b)x^3 + (ab + c + d)x^2 + (ad + bc)x + cd.$$

We obtain the polynomial

$$P(x) = x^4 + 8x^3 + 23x^2 + 28x + 12,$$

which, fortunately, has integer roots. We can find them by checking the divisors of 12. We obtain

$$P(x) = (x + 1)(x + 2)^2(x + 3).$$

The polynomial factors in two ways as a product of quadratic polynomials:

$$P(x) = (x^2 + 4x + 3)(x^2 + 4x + 4)$$

and

$$P(x) = (x^2 + 3x + 2)(x^2 + 5x + 6).$$

Hence the solutions of the system are $(4, 4, 3, 4)$, $(4, 4, 4, 3)$, $(3, 5, 2, 6)$, $(5, 3, 6, 2)$.

4.9 The Rearrangements Inequality

Problem 1.108 Let a, b, c be positive real numbers. Prove the inequality

$$\frac{a}{b+c} + \frac{b}{c+a} + \frac{c}{a+b} \geq \frac{3}{2}.$$

Solution For symmetry reasons, we can assume with no loss of generality that $a \leq b \leq c$. But then

$$\frac{1}{b+c} \leq \frac{1}{c+a} \leq \frac{1}{a+b},$$

and the rearrangements inequality gives

$$\frac{a}{b+c} + \frac{b}{c+a} + \frac{c}{a+b} \geq \frac{b}{b+c} + \frac{c}{c+a} + \frac{a}{a+b},$$

and also

$$\frac{a}{b+c} + \frac{b}{c+a} + \frac{c}{a+b} \geq \frac{c}{b+c} + \frac{a}{c+a} + \frac{b}{a+b}.$$

Adding the last two inequalities yields

$$2\left(\frac{a}{b+c} + \frac{b}{c+a} + \frac{c}{a+b}\right) \geq 3,$$

which is what we wanted to prove.

Problem 1.109 Let a, b, c be positive real numbers. Prove the inequality

$$\frac{a^3}{b^2+c^2} + \frac{b^3}{c^2+a^2} + \frac{c^3}{a^2+b^2} \geq \frac{a+b+c}{2}.$$

Solution Again, assume that $a \leq b \leq c$. This implies $a^2 \leq b^2 \leq c^2$, and it is not difficult to check that

$$\frac{a}{b^2+c^2} \leq \frac{b}{c^2+a^2} \leq \frac{c}{a^2+b^2}$$

also holds.

Applying the rearrangements inequality, we obtain

$$\frac{a^3}{b^2+c^2} + \frac{b^3}{c^2+a^2} + \frac{c^3}{a^2+b^2} \geq \frac{ab^2}{b^2+c^2} + \frac{bc^2}{c^2+a^2} + \frac{ca^2}{a^2+b^2},$$

and

$$\frac{a^3}{b^2+c^2} + \frac{b^3}{c^2+a^2} + \frac{c^3}{a^2+b^2} \geq \frac{ac^2}{b^2+c^2} + \frac{ba^2}{c^2+a^2} + \frac{cb^2}{a^2+b^2}.$$

Adding yields

$$2\left(\frac{a^3}{b^2+c^2}+\frac{b^3}{c^2+a^2}+\frac{c^3}{a^2+b^2}\right)\geq\frac{a(b^2+c^2)}{b^2+c^2}+\frac{b(c^2+a^2)}{c^2+a^2}+\frac{c(a^2+b^2)}{a^2+b^2}$$

$$=a+b+c,$$

as desired.

Problem 1.110 Let a,b,c be positive real numbers. Prove that

$$a+b+c\leq\frac{a^2+b^2}{2c}+\frac{b^2+c^2}{2a}+\frac{c^2+a^2}{2b}\leq\frac{a^3}{bc}+\frac{b^3}{ca}+\frac{c^3}{ab}.$$

Solution As usual, we assume that $a\leq b\leq c$. It follows that $a^2\leq b^2\leq c^2$, and $\frac{1}{a}\geq\frac{1}{b}\geq\frac{1}{c}$. Therefore,

$$a+b+c=a^2\cdot\frac{1}{a}+b^2\cdot\frac{1}{b}+c^2\cdot\frac{1}{c}\leq b^2\cdot\frac{1}{a}+c^2\cdot\frac{1}{b}+a^2\cdot\frac{1}{c},$$

and

$$a+b+c\leq c^2\cdot\frac{1}{a}+a^2\cdot\frac{1}{b}+b^2\cdot\frac{1}{c}.$$

Adding up and dividing by 2 yield

$$a+b+c\leq\frac{a^2+b^2}{2c}+\frac{b^2+c^2}{2a}+\frac{c^2+a^2}{2b}.$$

For the second inequality, observe that $a\leq b\leq c$ also implies $a^3\leq b^3\leq c^3$ and $\frac{1}{bc}\leq\frac{1}{ca}\leq\frac{1}{ab}$, hence we obtain

$$\frac{a^3}{bc}+\frac{b^3}{ca}+\frac{c^3}{ab}=a^3\cdot\frac{1}{bc}+b^3\cdot\frac{1}{ca}+c^3\cdot\frac{1}{ab}$$

$$\geq b^3\cdot\frac{1}{bc}+c^3\cdot\frac{1}{ca}+a^3\cdot\frac{1}{ab}$$

$$=\frac{b^2}{c}+\frac{c^2}{a}+\frac{a^2}{b},$$

and

$$\frac{a^3}{bc}+\frac{b^3}{ca}+\frac{c^3}{ab}\geq c^3\cdot\frac{1}{bc}+a^3\cdot\frac{1}{ca}+b^3\cdot\frac{1}{ab}$$

$$=\frac{c^2}{b}+\frac{a^2}{b}+\frac{b^2}{a}.$$

Adding up, we obtain

$$2\left(\frac{a^3}{bc} + \frac{b^3}{ca} + \frac{c^3}{ab}\right) \geq \frac{a^2+b^2}{c} + \frac{b^2+c^2}{a} + \frac{c^2+a^2}{b},$$

which proves the second inequality.

Problem 1.111 Let a, b, c be positive real numbers. Prove the inequality

$$\frac{a^2b(b-c)}{a+b} + \frac{b^2c(c-a)}{b+c} + \frac{c^2a(a-b)}{c+a} \geq 0.$$

Solution A first attempt is by brute force. Clearing out the denominators and reducing similar terms lead to the equivalent inequality

$$a^3b^3 + b^3c^3 + c^3a^3 \geq a^2bc^3 + b^2ca^3 + c^2ab^3.$$

Now, if we want to use the rearrangements inequality, we must assume some ordering of the numbers a, b, and c. Unlike the previous inequalities, this one is not symmetric with respect to a, b, c (switching a and b, for instance, lead to another inequality). Of course, we can assume that one of the numbers, say a, is the smallest one (that is because the inequality is invariant under cyclic permutations of a, b, c), but then we have to consider two cases: $a \leq b \leq c$ and $a \leq c \leq b$.

Let us assume that $a \leq b \leq c$. Then $ab \leq ac \leq bc$ and $a^2b^2 \leq a^2c^2 \leq b^2c^2$, hence

$$\begin{aligned}
a^3b^3 + a^3c^3 + b^3c^3 &= (ab)(a^2b^2) + (ac)(a^2c^2) + (bc)(b^2c^2) \\
&\geq (ac)(a^2b^2) + (bc)(a^2c^2) + (ab)(b^2c^2) \\
&= a^2bc^3 + b^2ca^3 + c^2ab^3,
\end{aligned}$$

and we are done. The case $a \leq c \leq b$ can be treated in a similar manner.

Observation We can prove our inequality with less computations. First, write it as

$$\frac{a^2b^2}{a+b} + \frac{b^2c^2}{b+c} + \frac{c^2a^2}{c+a} \geq \frac{a^2bc}{a+b} + \frac{b^2ca}{b+c} + \frac{c^2ab}{c+a},$$

and then divide both sides by abc:

$$\frac{ab}{c(a+b)} + \frac{bc}{a(b+c)} + \frac{ca}{b(c+a)} \geq \frac{a}{a+b} + \frac{b}{b+c} + \frac{c}{c+a}.$$

Assuming, for instance, $a \leq b \leq c$, we also have $ab \leq ac \leq bc$, and

$$\frac{1}{c(a+b)} \leq \frac{1}{a(b+c)} \leq \frac{1}{b(c+a)}.$$

It follows that

$$\frac{ab}{c(a+b)} + \frac{bc}{a(b+c)} + \frac{ca}{b(c+a)} \geq (ac)\frac{1}{c(a+b)} + (ab)\frac{1}{a(b+c)} + (bc)\frac{1}{b(c+a)}$$

$$= \frac{a}{a+b} + \frac{b}{b+c} + \frac{c}{c+a},$$

as desired. Analogously we deal with the case $a \leq c \leq b$.

Problem 1.112 Let $a_1 \leq a_2 \leq \cdots \leq a_n$ and $b_1 \leq b_2 \leq \cdots \leq b_n$ be two ordered sequences of real numbers. Prove Chebyshev's inequality

$$\frac{a_1 + a_2 + \cdots + a_n}{n} \cdot \frac{b_1 + b_2 + \cdots + b_n}{n} \leq \frac{a_1 b_1 + a_2 b_2 + \cdots + a_n b_n}{n}.$$

Solution The rearrangements inequality yields

$$a_1 b_1 + a_2 b_2 + \cdots + a_n b_n \geq a_1 b_2 + a_2 b_3 + \cdots + a_n b_1,$$

$$a_1 b_1 + a_2 b_2 + \cdots + a_n b_n \geq a_1 b_3 + a_2 b_4 + \cdots + a_n b_2,$$

$$\vdots$$

$$a_1 b_1 + a_2 b_2 + \cdots + a_n b_n \geq a_1 b_n + a_2 b_1 + \cdots + a_n b_{n-1},$$

$$a_1 b_1 + a_2 b_2 + \cdots + a_n b_n = a_1 b_1 + a_2 b_2 + \cdots + a_n b_n.$$

Adding up all and factoring the right-hand side lead to

$$n(a_1 b_1 + a_2 b_2 + \cdots + a_n b_n) \geq (a_1 + a_2 + \cdots + a_n)(b_1 + b_2 + \cdots + b_n),$$

and the conclusion follows immediately.

Problem 1.113 Let $a_1 \leq a_2 \leq \cdots \leq a_n$ and $b_1 \leq b_2 \leq \cdots \leq b_n$ be two ordered sequences of positive real numbers. Prove that

$$(a_1 + b_1)(a_2 + b_2) \cdots (a_n + b_n) \leq (a_1 + b_{\sigma(1)})(a_2 + b_{\sigma(2)}) \cdots (a_n + b_{\sigma(n)}),$$

for any permutation σ.

Solution We will use the same technique as in the proof of the rearrangements inequality. Consider an arbitrary permutation σ and the product

$$P(\sigma) = (a_1 + b_{\sigma(1)})(a_2 + b_{\sigma(2)}) \cdots (a_n + b_{\sigma(n)}).$$

If there exist $i < j$ such that $\sigma(i) > \sigma(j)$, consider the permutation σ' in which $\sigma(i)$ and $\sigma(j)$ are switched. We claim that $P(\sigma') \leq P(\sigma)$. Indeed, this is equivalent to

$$(a_i + b_{\sigma(j)})(a_j + b_{\sigma(i)}) \leq (a_i + b_{\sigma(i)})(a_j + b_{\sigma(j)}),$$

which simplifies to

$$(a_i - a_j)(b_{\sigma(i)} - b_{\sigma(j)}) \le 0,$$

clearly true, since $a_i \le a_j$ and $b_{\sigma(i)} \ge b_{\sigma(j)}$.

We repeat this procedure as long as we can find pairs (i, j), with $i < j$ and $\sigma(i) > \sigma(j)$. It is not difficult to see that, eventually, no such pair will exist. But in this case we have

$$\sigma(1) < \sigma(2) < \cdots < \sigma(n),$$

which clearly implies $\sigma(k) = k$, for all k. Thus, the product

$$(a_1 + b_1)(a_2 + b_2) \cdots (a_n + b_n)$$

is the minimal one.

Chapter 5
Geometry and Trigonometry

5.1 Geometric Inequalities

Problem 2.5 Let $ABCD$ be a convex quadrilateral. Prove that

$$\max(AB + CD, AD + BC) < AC + BD < AB + BC + CD + DA.$$

Solution Let O be the point of intersection of the diagonals AC and BD. We have $AO + OB > AB$ and $CO + OD > CD$; thus $AC + BD > AB + CD$. Similarly, $AO + OD > AD$ and $BO + OC > BC$; thus $AC + BD > AD + BC$. It follows that

$$\max(AB + CD, AD + BC) < AC + BD.$$

For the second inequality, note that $AC < AB + BC$ and $AC < AD + DC$; hence

$$AC < \frac{1}{2}(AB + BC + CD + DA).$$

Analogously,

$$BD < \frac{1}{2}(AB + BC + CD + DA),$$

and the result follows by adding these inequalities.

Problem 2.6 Let M be the midpoint of the segment AB. Prove that if O is an arbitrary point, then

$$|OA - OB| \le 2OM.$$

Solution Suppose that O does not lie on the line AB and let the point be O' the reflection of O across M (Fig. 5.1).

Because the segments AB and OO' have the same midpoint, the quadrilateral $AOBO'$ is a parallelogram, so $OA = BO'$. In the triangle OBO', we have

T. Andreescu, B. Enescu, *Mathematical Olympiad Treasures*,
DOI 10.1007/978-0-8176-8253-8_5, © Springer Science+Business Media, LLC 2011

Fig. 5.1

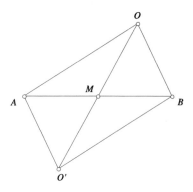

$OO' + OB > BO'$ and $OO' + BO' > OB$; hence $2OM > OA - OB$ and $2OM > OB - OA$.

We conclude that

$$2OM > |OA - OB|.$$

If O lies on the line AB, we introduce coordinates on the line AB such that M is the origin and the points A and B have coordinates -1 and 1, respectively. If x is the coordinate of the point O, we have to prove that

$$2|x| \geq \big||x + 1| - |x - 1|\big|.$$

Squaring both sides and rearranging terms, we find that the inequality is equivalent to $|x^2 - 1| \geq 1 - x^2$, which is obvious. We also deduce that the equality holds if and only if $1 - x^2 \geq 0$; that is, if and only if O lies on the segment AB.

Problem 2.7 Prove that in an arbitrary triangle, the sum of the lengths of the altitudes is less than the triangle's perimeter.

Solution It is not difficult to see that in a right triangle, the length of a leg is less than the length of the hypotenuse; this is an immediate consequence of the Pythagorean theorem, but it can also be proven by using simple inequalities.

Suppose ABC is a right triangle, with $\angle A = 90°$. We prove, for instance, that $BC > AB$. Let B' be the reflection of B across A. Then $\angle B'AC = 90°$ and triangles ABC and $AB'C$ are congruent. We have $BC + B'C > BB'$, hence $2BC > 2AB$. Returning to our problem, if AA', BB' and CC' are the altitudes of triangle ABC, then $AA' \leq AB$, $BB' \leq BC$, and $CC' \leq CA$. We add these inequalities to obtain

$$AA' + BB' + CC' < AB + BC + CA.$$

The inequality is strict because at most one of the angles of the triangle can be right.

Fig. 5.2

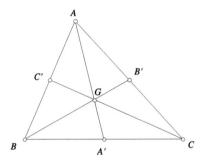

Problem 2.8 Denote by P the perimeter of triangle ABC. If M is a point in the interior of the triangle, prove that

$$\frac{1}{2}P < MA + MB + MC < P.$$

Solution In the triangle MAB, we have $MA + MB > AB$. Analogously, $MA + MC > AC$ and $MB + MC > BC$. The first inequality is obtained by adding these three inequalities. For the second one, we have seen in Problem 2.3 that $MB + MC < AB + AC$. Writing the similar inequalities and adding them up gives the desired result.

Problem 2.9 Prove that if A', B' and C' are the midpoints of the sides BC, CA and AB, respectively, then (Fig. 5.2)

$$\frac{3}{4}P < AA' + BB' + CC' < P.$$

Solution It is known that the medians of a triangle are concurrent at a point called the centroid of the triangle. Denote this point by G. Then

$$AA' + BB' + CC' > GA + GB + GC > \frac{1}{2}P,$$

which follows from the preceding problem by taking $M = G$, but this is not enough.

In order to obtain the required inequality, we have to use another property of the centroid: the point G divides each median in the ratio 1:2, that is, $GA = \frac{2}{3}AA'$, etc. Then, in triangle GBC, we have $GB + GC > BC$, or $\frac{2}{3}(BB' + CC') > BC$. If we add this with the other two similar inequalities, we get

$$\frac{4}{3}(AA' + BB' + CC') > P,$$

and hence the desired result. For the second inequality, just use Problem 2.2.

Problem 2.10 In the convex quadrilateral $ABCD$, we have

$$AB + BD \leq AC + CD.$$

Prove that $AB < AC$.

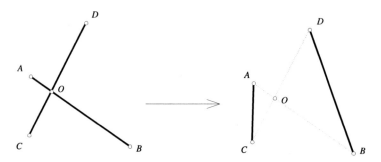

Fig. 5.3

Solution Let O be the point of intersection of the diagonals AC and BD. In triangles AOB and COD, we have $AB < AO + OB$, $CD < CO + OD$, and hence $AB + CD < AC + BD$. Adding this inequality with $AB + BD \leq AC + CD$, we get $AB < AC$.

Problem 2.11 Consider n red and n blue points in the plane, no three of them being collinear. Prove that one can connect each red point to a blue one with a segment such that no two segments intersect.

Solution There is a finite number of ways in which the n red points can be connected with the n blue ones. We choose the connection with the property that the sum of lengths of all connecting segments is minimal. We claim that in this case every two segments are disjoint. Indeed, suppose the points A and D are red, B and C are blue and the segments AB and CD intersect at O. Then we can replace the segments AB and CD by AC and BD (see Fig. 5.3). Clearly, $AC + BD < AB + CD$ (see Problem 2.5), and this contradicts the assumption that the sum of lengths of all connecting segments is minimal.

Problem 2.12 Let n be an odd positive integer. On some field, n gunmen are placed such that all pairwise distances between them are different. At a signal, every gunman takes out his gun and shoots the closest gunman. Prove that:

(a) at least one gunman survives;
(b) no gunman is shot more than five times;
(c) the trajectories of the bullets do not intersect.

Solution (a) Consider the gunmen closest to each other. They will shoot each other. If anybody else shoots at either of these two, a gunman will certainly survive (there are n bullets shot, so if someone ends up with more than one bullet in his body, someone else survives). If not, discard these two and repeat the reasoning for the $n - 2$ remaining gunmen. Because n is odd, we eventually reach the case $n = 3$, where the conclusion is obvious.

Fig. 5.4

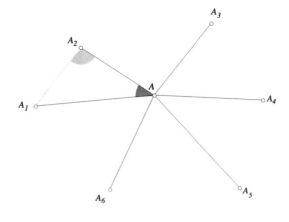

Fig. 5.5

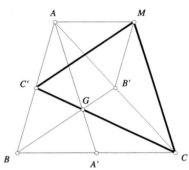

(b) Suppose a gunman A is shot by at least six other gunmen, denoted $A_1, \ldots, A_6$. One of the six angles centered at A is at most $60°$. Suppose $\angle A_1 A A_2$ is such an angle (see Fig. 5.4). In triangle $A_1 A A_2$, one of the other two angles, say A_2, is greater than $60°$. Then $AA_1 > A_1 A_2$, but since A_1 shot A and not A_2, $AA_1 < A_1 A_2$, which is a contradiction.

(c) Suppose A shoots B, C shoots D and $AB \cap CD = O$. Then $AB < AD$ and $CD < BC$. Thus, in the convex quadrilateral $ACBD$, $AB + CD < AD + BC$, a contradiction.

Problem 2.13 Prove that the medians of a given triangle can form a triangle.

Solution Let A', B' and C' be the midpoints of the sides BC, CA, and AB of the triangle ABC. We have $\overline{AA'} = \frac{1}{2}(\overline{AB} + \overline{AC})$. Adding this with the other two similar equalities, we obtain $\overline{AA'} + \overline{BB'} + \overline{CC'} = 0$. Since the three vectors are non-collinear, this shows that the segments AA', BB' and CC' can form a triangle.

Observation We can actually construct this triangle. Consider the point M such that B' is the midpoint of the segment $A'M$ (see Fig. 5.5). Then the quadrilateral $AA'CM$ is a parallelogram.

Fig. 5.6

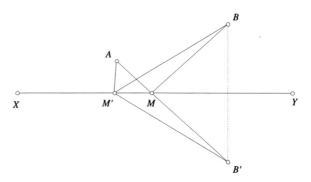

It is not difficult to see that in this case $BC'MB'$ is also a parallelogram, and the side lengths of triangle $CC'M$ are equal to the lengths of the medians AA', BB', and CC'.

Problem 2.14 Let A and B be two points situated on the same side of a line XY. Find the position of a point M on the line such that the sum $AM + MB$ is minimal.

Solution Let B' be the reflection of B across the line XY. We claim that M is the point of intersection of AB' and XY.

Indeed, let M' be another point on XY (see Fig. 5.6). By symmetry across XY, $MB' = MB$ and $M'B' = M'B$. Then

$$AM' + M'B = AM' + M'B' > AB' = AM + MB' = AM + MB.$$

Problem 2.15 Let ABC be an acute triangle. Find the positions of the points M, N, P on the sides BC, CA, AB, respectively, such that the perimeter of the triangle MNP is minimal.

Solution Let us fix M on the side BC and look for the positions of P and Q such that the perimeter of triangle MNP is minimal. Reflect M across AB and AC to obtain M' and M'', respectively.

If P' and Q' are points on the sides AB and AC, we have

$$MP' + P'Q' + Q'M = M'P' + P'Q' + Q'M''.$$

This sum is minimal when the points M', P', Q' and M'' are collinear, so P and Q are the points of intersection between $M'M''$ and the sides AB and AC (see Fig. 5.7). In this case, the perimeter of MPQ equals $M'M''$.

Now, the problem can be rephrased in the following way: find the point M on the side BC such that the length of $M'M''$ is minimal.

Observe that

$$\angle M'AB = \angle BAM, \quad \text{and} \quad \angle M''AC = \angle CAM,$$

Fig. 5.7

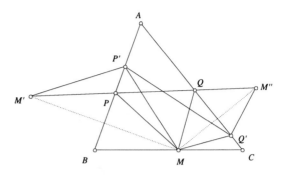

Fig. 5.8

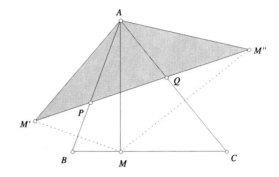

therefore $\angle M'AM'' = 2\angle BAC$. Moreover, $M'A = M''A = MA$. It follows that regardless of the position of M on BC, triangle $M'AM''$ is isosceles, with fixed angles.

All such triangles are similar to each other, so if $M'M''$ is minimal, the sides $M'A$ and $M''A$ are also minimal. Clearly, this happens when MA is minimal, i.e. when MA is an altitude of ABC (see Fig. 5.8). We conclude that the perimeter of MPQ is minimal when M, P, and Q are the feet of the altitudes of the triangle ABC (the so-called orthic triangle).

Problem 2.16 Seven real numbers are given in the interval $(1, 13)$. Prove that at least three of them are the lengths of a triangle's sides.

Solution Let the numbers be $a_1, a_2, \ldots, a_7$. We can assume that

$$a_1 \le a_2 \le \cdots \le a_7.$$

Suppose by way of contradiction that no three of them are the lengths of a triangle's sides. We have $a_1 + a_3 > a_2$ and $a_2 + a_3 > a_1$. Then $a_1 + a_2 \le a_3$, because if not, then there would exist a triangle with side lengths a_1, a_2, a_3.

We deduce $a_3 \ge 2$. Similarly, we have $a_4 \ge a_3 + a_2 \ge 3$, $a_5 \ge a_4 + a_3 \ge 5$, $a_6 \ge a_5 + a_4 \ge 8$ and, finally, $a_7 \ge a_6 + a_5 \ge 13$, which is a contradiction.

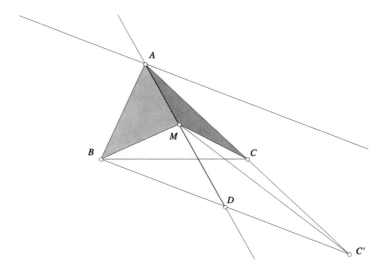

Fig. 5.9

5.2 An Interesting Locus

Problem 2.21 Find the locus of points M in plane of triangle ABC such that $[ABM] = 2[ACM]$.

Solution Let C' be the reflection of A across C. Then $[AMC'] = 2[AMC]$, so we have to determine the locus of points M for which $[ABM] = [AC'M]$ (Fig. 5.9).

It follows that the required locus is the union of the line AD (D being the midpoint of BC') and a parallel to BC' containing the point A.

Problem 2.22 Let D be a point on the side BC of triangle ABC and M a point on AD. Prove that

$$\frac{[ABM]}{[ACM]} = \frac{BD}{CD}$$

Deduce Ceva's theorem: if the segments AD, BE and CF are concurrent then

$$\frac{BD}{CD} \cdot \frac{CE}{AE} \cdot \frac{AF}{BF} = 1.$$

Solution Let A' be the projection of A onto BC (Fig. 5.10). Then $[ABD] = \frac{1}{2} AA' \cdot BD$, $[ACD] = \frac{1}{2} AA' \cdot CD$, hence

$$\frac{[ABD]}{[ACD]} = \frac{BD}{CD}.$$

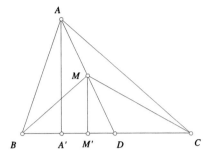 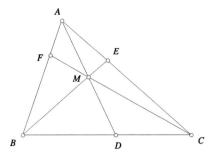

Fig. 5.10

Similarly, we obtain

$$\frac{[MBD]}{[MCD]} = \frac{BD}{CD}.$$

We deduce

$$\frac{BD}{CD} = \frac{[ABD] - [MBD]}{[ACD] - [MCD]} = \frac{[ABM]}{[ACM]},$$

as desired. For the second part, observe that

$$\frac{BD}{CD} \cdot \frac{CE}{AE} \cdot \frac{AF}{BF} = \frac{[ABM]}{[ACM]} \cdot \frac{[BCM]}{[ABM]} \cdot \frac{[ACM]}{[BCM]} = 1.$$

Problem 2.23 Let $ABCD$ be a convex quadrilateral and M a point in its interior such that

$$[MAB] = [MBC] = [MCD] = [MDA].$$

Prove that one of the diagonals of $ABCD$ passes through the midpoint of the other diagonal.

Solution Because $[MAB] = [MBC]$, the point M lies on the median of the triangle ABC. Similarly, we deduce that M lies on the median of the triangle ACD. If the two medians coincide, then they also coincide with BD, and then BD passes through the midpoint of AC.

   If not, then M must lie on AC. Since $[MAB] = [MAD]$, AC bisects BD, as desired.

Problem 2.24 Let $ABCD$ be a convex quadrilateral. Find the locus of points M in its interior such that

$$[MAB] = 2[MCD].$$

Solution We apply the same method as in Problem 2.17 (see Fig. 5.12).

   We have to find the locus of points M such that $[MXT] = 2[MYT]$. As we have seen in Problem 2.21, if Y' is a point on YC such that $TY = YY'$, then the

Fig. 5.11

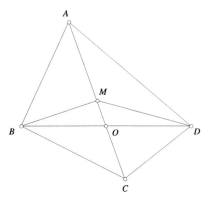

Fig. 5.12

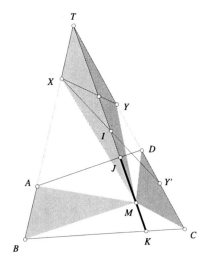

locus of M is the set of points in the interior of the quadrilateral $ABCD$ located on the line TI, where I is the midpoint of XY'. Therefore, the required locus is the segment JK.

Problem 2.25 Let $ABCD$ be a convex quadrilateral and let $k > 0$ be a real number. Find the locus of points M in its interior such that

$$[MAB] + 2[MCD] = k.$$

Solution Combining Problems 2.19 and 2.21 and using the construction above, we deduce that the area of $XY'M$ is constant (Fig. 5.13).

It results that the locus is (according to the value of k) either a segment KL parallel to XY', or a point, or the empty set.

Problem 2.26 Let d, d' be two non-parallel lines in the plane and let $k > 0$. Find the locus of points the sum of whose distances to d and d' is equal to k.

Fig. 5.13

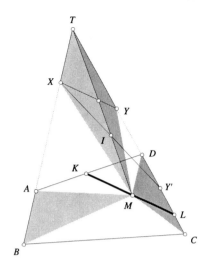

Solution Let O be the point of intersection between d and d'. We consider the points A, B, C, D, as in Fig. 5.14, such that $OA = OB = OC = OD = a > 0$. Suppose M lies in the interior of the angle AOB.

If M_A and M_B are the projections of M onto OA and OB, we have

$$MM_A + MM_B = k.$$

Multiplying by $a/2$, we obtain

$$[MOA] + [MOB] = \frac{ka}{2}.$$

But

$$[MOA] + [MOB] = [MAOB] = [OAB] + [MAB].$$

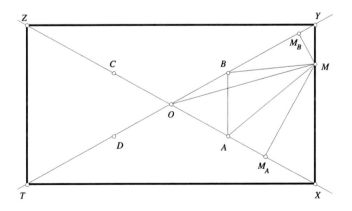

Fig. 5.14

Fig. 5.15

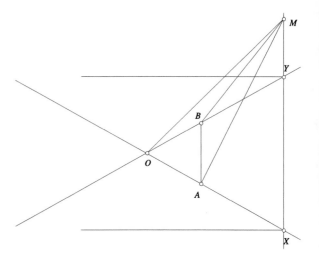

It follows that the area of MAB is constant, and hence the locus of M in the interior of $\angle AOB$ is a segment XY parallel to AB.

Considering the other three possible locations of M, we deduce that the locus is a rectangle $XYZT$ whose diagonals are d and d'.

Observation It is worth mentioning that if M lies, for instance, on the line XY but not in the interior of the segment XY, another equality occurs.

If M is on the half-line XY as in the figure above (Fig. 5.15), we have

$$[MOA] - [MOB] = [OAB] + [MAB],$$

so we deduce that for those positions of M, the difference of the distances to d and d' equals k.

Problem 2.27 Let $ABCD$ be a convex quadrilateral and let E and F be the points of intersections of the lines AB, CD and AD, BC, respectively. Prove that the midpoints of the segments AC, BD, and EF are collinear.

Solution Let P, Q, and R be the midpoints of AC, BD, and EF (Fig. 5.16). Denote by S the area of $ABCD$. As we have seen, the locus of the points M in the interior of $ABCD$ for which

$$[MAB] + [MCD] = \frac{1}{2}S$$

is a segment. We see that P and Q belong to this segment. Indeed,

$$[PAB] + [PCD] = \frac{1}{2}[ABC] + \frac{1}{2}[ACD] = \frac{1}{2}S,$$

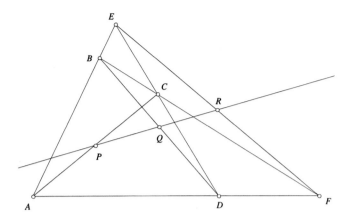

Fig. 5.16

Fig. 5.17

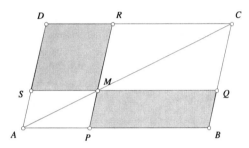

and

$$[QAD] + [QCD] = \frac{1}{2}[ABD] + \frac{1}{2}[BCD] = \frac{1}{2}S.$$

Now, we have $[RAB] = \frac{1}{2}[FAB]$, since the distance from F to AB is twice the distance from R to AB. Similarly, $[RCD] = \frac{1}{2}[FCD]$. We obtain

$$[RAB] - [RCD] = \frac{1}{2}([FAB] - [FCD]) = \frac{1}{2}S.$$

Taking into account the observation in the solution to Problem 3, it follows that P, Q and R are collinear.

Observation Another proof of the assertion in the problem can be obtained by using a simple (yet useful) lemma. We leave the proof as an exercise to the reader.

Lemma *Suppose that through the point M lying in the interior of the parallelogram $ABCD$, two parallels to AB and AD are drawn, intersecting the sides of $ABCD$ at the points P, Q, R, S (see Fig. 5.17).*
Then M lies on the diagonal AC if and only if $[MRDS] = [MPBQ]$.

Fig. 5.18

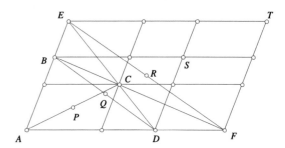

Returning to our problem, let us draw through the points B, C, D, E, and F parallels to AB and AD (see Fig. 5.18). It is not difficult to see that the points P, Q, and R are the images of C, S, and T through a homothety of center A and ratio 1/2, therefore it suffices to prove that C, S and T are collinear.

Now we apply the lemma, using the fact that C lies on both ED and BF to see that the shaded parallelograms in Fig. 5.19 have equal area.

Hence by subtracting we see that the shaded parallelograms in Fig. 5.20 have the same area and we obtain that S lies on CT.

Problem 2.28 In the interior of a quadrilateral $ABCD$, consider a variable point P. Prove that if the sum of distances from P to the sides is constant, then $ABCD$ is a parallelogram.

Solution Consider the points X, Y, Z, and T on the sides AB, BC, CD, and DA, respectively, such that

$$AX = AT = CY = CZ = a.$$

Let x, y, z, and t be the distances from M to the quadrilateral's sides. If $x+y+z+t$ is constant, then the expression

$$\frac{1}{2}a(x+y+z+t) = [MAX] + [MCY] + [MCZ] + [MAT]$$

$$= [ATX] + [CYZ] + [MXT] + [MYZ]$$

is constant, as well. Because $[ATX]$ and $[CYZ]$ do not depend of the position of M, it follows that the sum $[MXT] + [MYZ]$ is constant (Fig. 5.21).

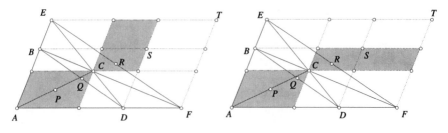

Fig. 5.19

Fig. 5.20

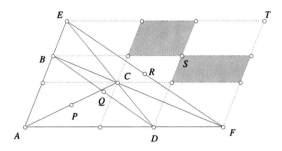

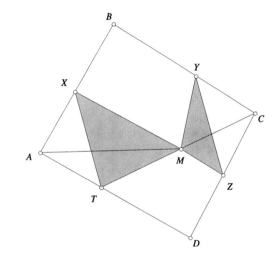

Fig. 5.21

As we have seen, if XT and YZ are not parallel, the locus of points M satisfying the above condition is a segment, and not all the interior points of the quadrilateral $XYZT$. It follows that XT is parallel to YZ, hence the angle bisectors of XAT and YCZ are parallel. From here we deduce that $\angle B = \angle D$. A similar argument yields $\angle A = \angle C$, hence $ABCD$ is a parallelogram.

Observation More generally, if $A_1 A_2 \ldots A_n$ is a convex polygon, then the sum of the distances from a variable interior point to its sides is constant if and only if the following equality holds:

$$\frac{1}{A_1 A_2} \cdot \overline{A_1 A_2} + \frac{1}{A_2 A_3} \cdot \overline{A_2 A_3} + \cdots + \frac{1}{A_n A_1} \cdot \overline{A_n A_1} = 0.$$

5.3 Cyclic Quads

Problem 2.32 Let D, E, and F be the feet of the altitudes of the triangle ABC. Prove that the altitudes of ABC are the angle bisectors of the triangle DEF (Fig. 5.22).

Fig. 5.22

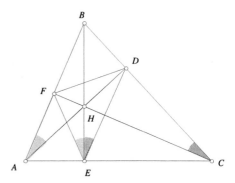

Solution We can see that the quadrilateral $AFHE$ is cyclic. Indeed, the angles $\angle AFH$ and $\angle AEH$ are right angles; thus $AFHE$ is inscribed in the circle of diameter AH. It follows that $\angle FAH = \angle FEH$.

Similarly, in the quadrilateral $CDHE$ we have $\angle DCH = \angle DEH$. But in the right triangles ABD and CBF, the angles $\angle FAH$ and $\angle DCH$ equal $90° - \angle B$, hence $\angle FAH = \angle DCH$. It follows that $\angle FEH = \angle DEH$; that is, BE is the angle bisector of $\angle DEF$.

Problem 2.33 Let $ABCD$ be a convex quadrilateral. Prove that

$$AB \cdot CD + AD \cdot BC = AC \cdot BD$$

if and only if $ABCD$ is cyclic (Ptolemy's theorem).

Solution Consider the point A' such that triangles ABD and $A'BC$ are similar (Fig. 5.23). Then

$$\frac{AD}{BD} = \frac{A'C}{BC},$$

hence

$$AD \cdot BC = A'C \cdot BD. \tag{5.1}$$

We also have

$$\frac{AB}{A'B} = \frac{BD}{BC}$$

and since $\angle ABA' = \angle DBC$, it follows that triangles ABA' and DBC are similar, too.

We deduce that

$$\frac{AB}{AA'} = \frac{BD}{DC},$$

hence

$$AB \cdot DC = AA' \cdot BD. \tag{5.2}$$

Fig. 5.23

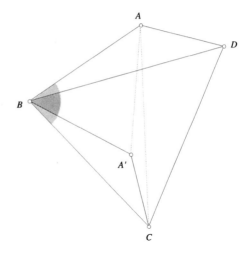

Adding up (5.1) and (5.2) we obtain

$$AD \cdot BC + AB \cdot CD = (A'C + AA') \cdot BD.$$

But in triangle $AA'C$, $A'C + AA' \geq AC$, hence

$$AD \cdot BC + AB \cdot CD \geq AC \cdot BD.$$

This inequality holds in every convex quadrilateral and it turns into an equality if and only if triangle $AA'C$ is degenerate, that is, the point A' lies on AC. This happens if and only if $\angle A'CB = \angle ACB$. But since $\angle A'CB = \angle ADB$, the equality holds if and only if $\angle ADB = \angle ACB$ and this is the condition that $ABCD$ is cyclic.

Problem 2.34 Let A', B' and C' be points in the interior of the sides BC, CA and AB of the triangle ABC. Prove that the circumcircles of the triangles $AB'C'$, $BA'C'$ and $CA'B'$ have a common point.

Solution Let M be the point of intersection of the circumcircles of triangles $AB'C'$ and $BC'A'$. Because the quadrilateral $AB'MC'$ is cyclic, $\angle MC'A = \angle MB'C$. Similarly, since $BC'MA'$ is cyclic, $\angle MA'B = \angle MC'A$. It follows that $\angle MB'C = \angle MA'B$, hence $MA'CB'$ is also cyclic. This means that the circumcircle of triangle $CA'B'$ passes through M (Fig. 5.24).

Observation The property holds even if the points A', B' and C' are collinear (clearly, in this case the points are not all three in the interior of the triangle's sides). Suppose that A' and B' are in the interior of the sides BC and AC and C' is on the line AB such that the three points are collinear. Let M be the point of intersection between the circumcircles of triangles $AB'C'$ and $BC'A'$. Then $\angle AC'M = \angle MB'C$ ($AB'MC'$ is cyclic) and $\angle AC'M = \angle MA'C$ ($BA'MC'$ is cyclic). It follows that $\angle MB'C = \angle MA'C$, thus $MB'A'C$ is also cyclic.

Fig. 5.24

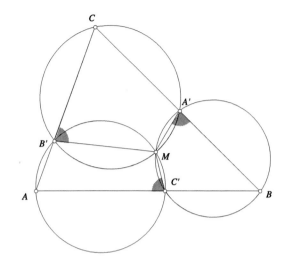

Fig. 5.25

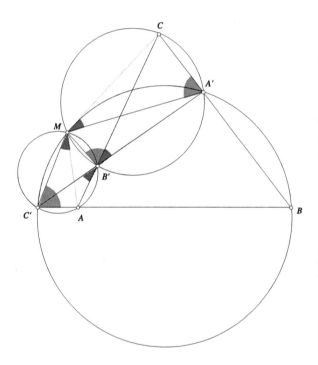

Moreover, in the cyclic quadrilateral $AB'MC'$, we have $\angle AMC' = \angle AB'C'$. Similarly, $\angle A'MC = \angle A'B'C$. Because $\angle AB'C' = \angle A'B'C$, it follows that $\angle AMC' = \angle A'MC$, hence $\angle A'MC' = \angle AMC$. But $\angle A'MC' + \angle ABC = 180°$ ($A'BC'M$ is cyclic) so $\angle AMC + \angle ABC = 180°$. This means that $ABCM$ is cyclic, hence the circumcircle of triangle ABC also passes through M (Fig. 5.25).

Problem 2.35 Let $ABCD$ be a cyclic quadrilateral. Prove that the orthocenters of the triangles ABC, BCD, CDA and DAB are the vertices of a quadrilateral congruent to $ABCD$ and prove that the centroids of the same triangles are the vertices of a cyclic quadrilateral.

Solution We first prove a very useful result.

Lemma *Let H be the orthocenter of a triangle ABC and O its circumcenter. Then the vectors $\overline{OA}, \overline{OB}, \overline{OC}$ and $\overline{OH}$ satisfy the following equality:*

$$\overline{OA} + \overline{OB} + \overline{OC} = \overline{OH}.$$

Proof Let A' be a point on the circumcircle such that AA' is a diameter. Then the quadrilateral $A'BHC$ is a parallelogram. Indeed, since AA' is a diameter, $\angle A'CA = 90°$, so that $A'C \perp AC$. But we also have $BH \perp AC$, thus $A'C$ and BH are parallel. Similarly it follows that $A'B$ and CH are parallel, hence $A'BHC$ is a parallelogram.
   We have

$$\overline{HC} + \overline{HB} = \overline{HA'}$$

and

$$\overline{HA'} + \overline{HA} = 2\overline{HO}$$

(O is the midpoint of AA'), so that

$$\overline{HA} + \overline{HB} + \overline{HC} = 2\overline{HO}.$$

But then

$$\overline{OA} + \overline{OB} + \overline{OC} = \overline{OH} + \overline{HA} + \overline{OH} + \overline{HB} + \overline{OH} + \overline{HC}$$
$$= 3\overline{OH} + 2\overline{HO} = \overline{OH},$$

as desired (Fig. 5.26).
   It is worth mentioning that if G is the centroid of triangle ABC, then

$$\overline{OA} + \overline{OB} + \overline{OC} = 3\overline{OG},$$

hence $\overline{OH} = 3\overline{OG}$. This means that the points O, G and H are collinear and $OH = 3OG$.   □

   Returning to the problem, let O be circumcenter of the quadrilateral $ABCD$ and H_A, H_B, H_C, H_D the orthocenters of triangles BCD, CDA, DAB, ABC, respectively. Using the lemma, we have

$$\overline{H_A H_B} = \overline{OH_B} - \overline{OH_A} = \left(\overline{OC} + \overline{OD} + \overline{OA}\right) - \left(\overline{OB} + \overline{OC} + \overline{OD}\right)$$
$$= \overline{OA} - \overline{OB} = \overline{BA},$$

Fig. 5.26

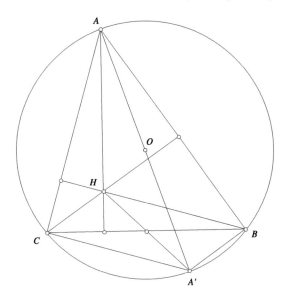

Fig. 5.27

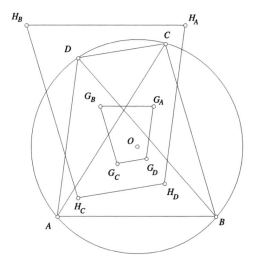

hence the segments $H_A H_B$ and AB are equal and parallel. We conclude that the quadrilaterals $ABCD$ and $H_A H_B H_C H_D$ are equal.

For the second claim, denoting by G_A, G_B, G_C and G_D the centroids, it follows that from the observation above that $G_A G_B G_C G_D$ is obtained from $H_A H_B H_C H_D$ by a homothety of center O and ratio $1/3$. Because $H_A H_B H_C H_D$ is cyclic, so is $G_A G_B G_C G_D$ (Fig. 5.27).

Problem 2.36 Let K, L, M, N be the midpoints of the sides AB, BC, CD, DA, respectively, of a cyclic quadrilateral $ABCD$. Prove that the orthocenters of triangles AKN, BKL, CLM, DMN are the vertices of a parallelogram.

Fig. 5.28

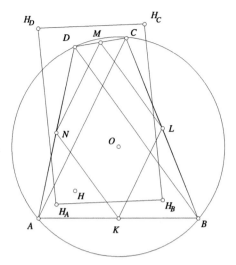

Solution Denote by H_A, H_B, H_C, H_D the orthocenters of triangles AKN, BKL, CLM and DMN. If O is the circumcenter of $ABCD$ and H is the orthocenter of ABD, using the lemma from Problem 2.35, we have

$$\overline{OH} = \overline{OA} + \overline{OB} + \overline{OD}.$$

Because K and N are the midpoints of AB and AD, triangle AKN is the image of ABD through a homothety of center A and ratio $1/2$. Thus, $\overline{AH_A} = \frac{1}{2}\overline{AH}$. Then we have

$$\overline{OH_A} = \overline{OA} + \overline{AH_A} = \overline{OA} + \frac{1}{2}\overline{AH} = \overline{OA} + \frac{1}{2}\left(\overline{OH} - \overline{OA}\right)$$

$$= \overline{OA} + \frac{1}{2}\left(\overline{OB} + \overline{OD}\right).$$

Similarly,

$$\overline{OH_C} = \overline{OC} + \frac{1}{2}\left(\overline{OB} + \overline{OD}\right),$$

and by adding these equalities, we obtain

$$\overline{OH_A} + \overline{OH_C} = \overline{OA} + \overline{OB} + \overline{OC} + \overline{OD}.$$

From the symmetry of the right-hand side, it follows that

$$\overline{OH_A} + \overline{OH_C} = \overline{OH_B} + \overline{OH_D},$$

and this equality implies that the segments $H_A H_C$ and $H_B H_D$ have the same midpoint, hence $H_A H_B H_C H_D$ is a parallelogram (Fig. 5.28).

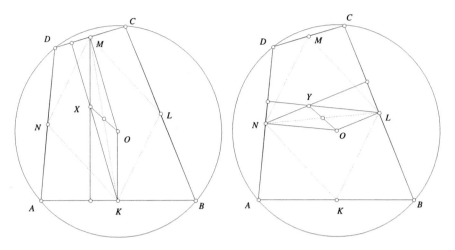

Fig. 5.29

Problem 2.37 Prove that the perpendiculars dropped from the midpoints of the sides of a cyclic quadrilateral to the opposite sides are concurrent.

Solution Let K and M be the midpoints of AB and CD and let X be the intersection point of the perpendiculars dropped from these points to the opposite sides. Let O be the circumcenter of $ABCD$. Because $OK \perp AB$ and $OM \perp CD$, the quadrilateral $OKXM$ is a parallelogram, hence the midpoint of OX coincides with the midpoint of KM.

Let L and N the midpoints of BC and AD and Y the intersection point of the perpendiculars dropped from these points to the opposite sides. It results that the midpoint of OY coincides with the midpoint of LN. It is not difficult to see that $KLMN$ is a parallelogram, hence KM and LN have the same midpoint. It follows that $X = Y$ and the claim is proved (Fig. 5.29).

Problem 2.38 In the convex quadrilateral $ABCD$ the diagonals AC and BD intersect at O and are perpendicular. Prove that projections of O on the quadrilateral's sides are the vertices of a cyclic quadrilateral.

Solution Let K, L, M, N be the projections of O on the sides AB, BC, CD, DA, respectively. The quadrilateral $AKON$ is cyclic, hence $\angle OAN = \angle OKN$ (Fig. 5.30).

In the same way we obtain $\angle OBL = \angle OKL$, $\angle ODN = \angle OMN$, $\angle OCL = \angle OML$. If we add these equalities, it follows that

$$\angle LKN + \angle LMN = \angle OAD + \angle ODA + \angle OBC + OCB = 90° + 90° = 180°,$$

hence $KLMN$ is cyclic.

Fig. 5.30

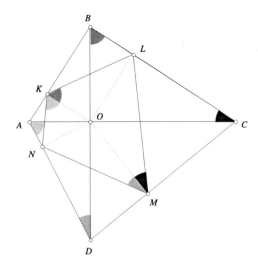

5.4   Equiangular Polygons

Problem 2.42 Let $ABCDE$ be an equiangular pentagon whose side lengths are rational numbers. Prove that the pentagon is regular.

Solution Let M and N be the intersection points of the line AE with BC and CD, respectively.

Because $ABCDE$ is equiangular, triangles AMB and DNE are isosceles, with $\angle M = \angle N = 36°$, therefore triangle CMN is also isosceles and $CM = CN$. It follows that $BC + BM = CD + DN$. But $BM = \frac{AB}{2\cos 72°}$ and $DN = \frac{DE}{2\cos 72°}$, hence $BC - CD = \frac{DE - AB}{2\cos 72°}$. If $DE \neq AB$ it follows that $\cos 72°$ is a rational number, which is a contradiction, since $\cos 72° = \frac{\sqrt{5}-1}{4}$. The conclusion is that $DE = AB$ and, by similar arguments, all the pentagon's sides have equal lengths (Fig. 5.31).

Observation An alternate solution is given in the next problem.

Problem 2.43 Prove that p is a prime number if and only if every equiangular polygon with p sides of rational lengths is regular.

Solution The proof uses some advanced knowledge in the algebra of polynomials. Suppose p is a prime number and let the rational numbers $a_1, a_2, \ldots, a_p$ be the side lengths of an equiangular polygon. We have seen that

$$\zeta = \cos \frac{2\pi}{p} + i \sin \frac{2\pi}{p}$$

is a root of the polynomial

$$P(X) = a_1 + a_2 X + \cdots + a_p X^{p-1}.$$

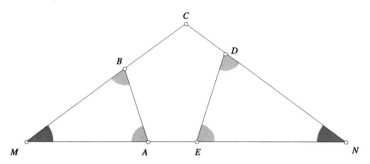

Fig. 5.31

On the other hand, ζ is also a root of the polynomial

$$Q(X) = 1 + X + X^2 + \cdots + X^{p-1}.$$

Because the two polynomials share a common root, their greatest common divisor must be a non-constant polynomial with rational coefficients.

Since Q canot be factorized as a product of two non-constant polynomials with rational coefficients (to prove that, one applies the Eisenstein criterion to the polynomial $Q(X + 1)$), it follows that for some constant c we have $P = cQ$, hence $a_1 = a_2 = \cdots = a_p$.

Conversely, suppose p is not a prime number and let $p = mn$, for some positive integers $m, n > 1$. It results that ζ^n is a root of order m of the unity, hence $1 + \zeta^n + \zeta^{2n} + \cdots + \zeta^{(m-1)n} = 0$. If we add this equality to $1 + \zeta + \zeta^2 + \zeta^3 + \cdots + \zeta^{p-1} = 0$, we deduce that ζ is the root of a polynomial of degree $p - 1$, with some coefficients equal to 1 and the others equal to 2. This means that there exists an equiangular polygon with p sides, some of length 1 and the rest of length 2. Because such a polygon is not regular, our claim is proved.

Observation We can examine, for instance, the case $p = 6 = 2 \cdot 3$. If

$$\zeta = \cos \frac{\pi}{3} + i \sin \frac{\pi}{3}$$

then

$$1 + \zeta^2 + \zeta^4 = 0$$

and

$$1 + \zeta + \zeta^2 + \zeta^3 + \zeta^4 + \zeta^5 = 0,$$

hence ζ is a root of the polynomial

$$2 + X + 2X^2 + X^3 + 2X^4 + X^5.$$

This means that there exists an equiangular hexagon with side lengths 2, 1, 2, 1, 2, 1, as can be seen in Fig. 5.32.

Fig. 5.32

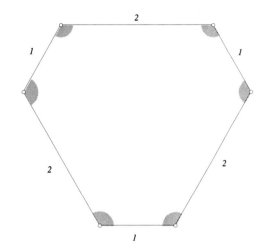

Fig. 5.33

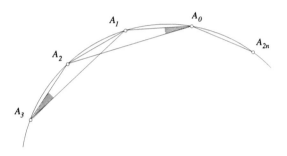

Problem 2.44 An equiangular polygon with an odd number of sides is inscribed in a circle. Prove that the polygon is regular.

Solution Let $A_0 A_1 \ldots A_{2n}$ be the polygon (Fig. 5.33). Then triangles $A_0 A_1 A_2$ and $A_1 A_2 A_3$ are congruent. Indeed, $A_1 A_2$ is a common side, the angles $\angle A_0 A_1 A_2$ and $\angle A_1 A_2 A_3$ are congruent since the polygon is equiangular, and the angles $\angle A_1 A_0 A_2$ and $\angle A_1 A_3 A_2$ are also congruent (their measure is $\frac{1}{2} \overset{\frown}{A_1 A_2}$). It follows that $A_0 A_1 = A_2 A_3$.

In the same way we obtain

$$A_2 A_3 = A_4 A_5 = \cdots = A_{2n-2} A_{2n-1} = A_{2n} A_0 = A_1 A_2 = \cdots = A_{2n-1} A_{2n},$$

hence the polygon is regular.

Problem 2.45 Let $a_1, a_2, \ldots, a_n$ be the side lengths of an equiangular polygon. Prove that if $a_1 \geq a_2 \geq \cdots \geq a_n$, then the polygon is regular.

Solution The first approach is geometrical. We examine two cases: n odd and n even. If n is odd, say $n = 2k + 1$, consider the angle bisector of $\angle A_{2k+1} A_1 A_2$. It

Fig. 5.34

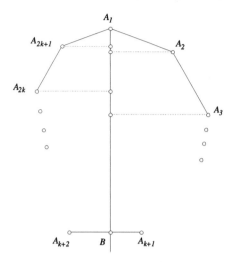

is not difficult to see that it is perpendicular to the side $A_{k+1}A_{k+2}$. Project all the sides of the polygon on this line. If we denote by x_i the length of the projection of the side $A_i A_{i+1}$ (with the usual convention $A_{2k+2} = A_1$), then

$$x_1 + x_2 + \cdots + x_k = x_{k+2} + x_{k+3} + \cdots + x_{2k+1} = A_1 B$$

(see Fig. 5.34). On the other hand, the angle between $A_i A_{i+1}$ and $A_1 B$ is equal to the angle between $A_{2k+2-i} A_{2k+3-i}$ and $A_1 B$, thus $x_i \geq x_{2k+2-i}$, for all $1 \leq i \leq k$. It follows that the above equality can be reached only if the sides of the polygon are equal.

A similar argument works in the case when n is even.

The second approach is algebraic. Let

$$\varepsilon = \cos \frac{2\pi}{n} + i \sin \frac{2\pi}{n}$$

be a primitive root of the unity. Then ε is a root of the polynomial

$$P(X) = a_1 + a_2 X + \cdots + a_n X^{n-1}.$$

The conclusion is obtained from the following:

Lemma *Let* $P(X) = a_1 + a_2 X + \cdots + a_n X^{n-1}$, *where* $a_1 \geq a_2 \geq \cdots \geq a_n > 0$. *If* α *is a root of* P, *then* $|\alpha| \geq 1$, *and* $|\alpha| = 1$ *only if* $a_1 = a_2 = \cdots = a_n$.

Proof We have

$$a_1 + a_2\alpha + \cdots + a_n\alpha^{n-1} = 0.$$

If we multiply this equality with $\alpha - 1$, we obtain

$$-a_1 + \alpha(a_1 - a_2) + \alpha^2(a_2 - a_3) + \cdots + \alpha^{n-1}(a_{n-1} - a_n) + a_n\alpha^n = 0,$$

Fig. 5.35

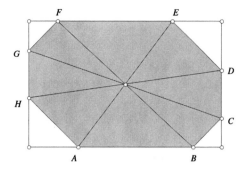

or, equivalently,

$$a_1 = \alpha(a_1 - a_2) + \alpha^2(a_2 - a_3) + \cdots + \alpha^{n-1}(a_{n-1} - a_n) + a_n\alpha^n.$$

Now, suppose that $|\alpha| \le 1$. It results

$$a_1 = \left| \alpha(a_1 - a_2) + \alpha^2(a_2 - a_3) + \cdots + \alpha^{n-1}(a_{n-1} - a_n) + a_n\alpha^n \right|$$
$$\le |\alpha|(a_1 - a_2) + |\alpha|^2(a_2 - a_3) + \cdots + |\alpha|^{n-1}(a_{n-1} - a_n) + a_n|\alpha|^n$$
$$\le (a_1 - a_2) + (a_2 - a_3) + \cdots + (a_{n-1} - a_n) + a_n = a_1.$$

Consequently, all inequalities must be equalities. Because $\alpha \notin \mathbf{R}$, this is possible only if $a_1 = a_2 = \cdots = a_n$, hence the polygon is regular. $\qquad \square$

Problem 2.46 The side lengths of an equiangular octagon are rational numbers. Prove that the octagon has a symmetry center.

Solution The angles of an equiangular octagon are equal to $135°$, thus, the lines containing the segments AB, CD, EF and GH determine a rectangle (Fig. 5.35).

Because the opposite sides of this rectangle are equal, we obtain

$$AB + \frac{\sqrt{2}}{2}(AH + BC) = EF + \frac{\sqrt{2}}{2}(DE + FG),$$

or, equivalently,

$$AB - EF = \frac{\sqrt{2}}{2}(DE + FG - AH - BC).$$

Because the side lengths of the octagon are rational numbers, the above equality can hold if and only if

$$AB - EF = DE + FG - AH - BC = 0.$$

In a similar way, we obtain

$$CD - GH = FG + AH - DE - BC = 0.$$

Fig. 5.36

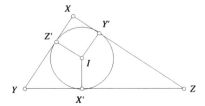

From these equalities it follows that $AB = EF$, $CD = GH$, $BC = FG$ and $DE = AH$, so the opposite sides of the octagon are equal and parallel. It follows that the quadrilaterals $ABEF$, $BCFG$, $CDGH$ and $DEHA$ are parallelograms, hence the midpoints of the segments AE, BF, CG, DH and DH coincide. Obviously, this common point is a symmetry center of the octagon.

5.5   More on Equilateral Triangles

Problem 2.49 Let M be a point in the interior of the equilateral triangle ABC and let A', B', C' be its projections onto the sides BC, CA and AB, respectively. Prove that the sum of lengths of the inradii of triangles MAC', MBA' and MCB' equals the sum of lengths of the inradii of triangles MAB', MBC' and MCA'.

Solution We start by proving an additional result.

Lemma *Let XYZ be a right triangle, with $\angle X = 90°$ and side lengths x, y, z, respectively. If r is the length of the inradius of XYZ, then $r = \frac{1}{2}(y + z - x)$.*

Proof Let X', Y', Z' the tangency points of the incircle with triangle's sides (Fig. 5.36). Then $r = XY' = XZ'$. Denote by $s = ZY' = ZX'$ and $t = YX' = YZ'$.
    Then $r + s = y$, $r + t = z$ and $s + t = x$. Solving for r, we obtain $r = \frac{1}{2}(y + z - x)$.   □

Applying the lemma in our problem, we have to prove that

$$(AC' + MC' - MA) + (BA' + MA' - MB) + (CB' + MB' - MC)$$
$$= (AB' + MB' - MA) + (BC' + MC' - MB) + (CA' + MA' - MC),$$

which is equivalent to

$$AB' + BC' + CA' = AC' + BA' + CB'.$$

To prove this equality, we draw through M parallels to triangle's sides, as in Fig. 5.37. Then $AB' = AB''' + B'''B'$, but $AB''' = BA''$ and $B'''B' = B'B''$. Writing the similar equalities, the result follows by adding them.   □

Fig. 5.37

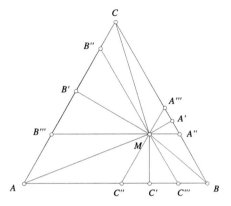

Observation Another way to deal with the last part of the problem is to use the Pythagorean theorem. We have $A'M^2 = BM^2 - A'B^2 = CM^2 - A'C^2$, hence $BM^2 - CM^2 = A'B^2 - A'C^2 = (A'B - A'C) \cdot a$, where a is the side length of the triangle. Writing the other two similar equalities and adding them we get the result.

Problem 2.50 Let I be the incenter of triangle ABC. It is known that for every point $M \in (AB)$, one can find the points $N \in (BC)$ and $P \in (AC)$ such that I is the centroid of triangle MNP. Prove that ABC is an equilateral triangle.

Solution Let a, b, c be the lengths of triangle's sides and let m, n, p be real numbers such that $\overline{AM} = m\overline{AB}$, $\overline{BN} = n\overline{BC}$ and $\overline{CP} = p\overline{CA}$. If G and G' are the centroids of triangles ABC and MNP, then

$$\overline{GG'} = \frac{1}{3}\left(\overline{GM} + \overline{GN} + \overline{GP}\right) = \frac{1}{3}\left(\overline{GA} + \overline{AM} + \overline{GB} + \overline{BN} + \overline{GC} + \overline{CP}\right)$$

$$= \frac{1}{3}\left(\overline{AM} + \overline{BN} + \overline{CP}\right) = \frac{1}{3}\left(m\overline{AB} + n\overline{BC} + p\overline{CA}\right)$$

$$= \frac{m-p}{3}\overline{AB} + \frac{n-p}{3}\overline{BC}.$$

Because I is the incenter of ABC, we have

$$\overline{GI} = \frac{a\overline{GA} + b\overline{GB} + c\overline{GC}}{a+b+c} = \frac{b+c-2a}{3(a+b+c)}\overline{AB} + \frac{2c-a-b}{3(a+b+c)}\overline{BC}.$$

It results that $G' = I$ if an only if

$$\begin{cases} m - p = \frac{b+c-2a}{a+b+c}, \\ n - p = \frac{2c-a-b}{a+b+c} \end{cases}$$

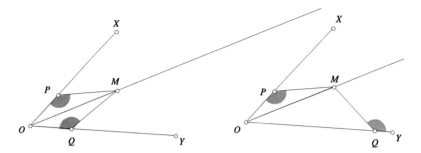

Fig. 5.38

or, equivalently

$$\begin{cases} n = m + 1 - \frac{3b}{a+b+c}, \\ p = m - 1 + \frac{3a}{a+b+c}. \end{cases} \qquad (*)$$

Therefore, for any point $M \in (AB)$, one can find the points $N \in BC$ and $P \in CA$ such that the centroid of MNP is the point I. The problem is that N and P must lie in the interior of the segments (BC) and (CA), respectively. This is equivalent to the following: for every $m \in (0, 1)$, the numbers n and p given by $(*)$ also belong to the interval $(0, 1)$. It is not difficult to see that this happens if and only if

$$1 - \frac{3b}{a+b+c} = -1 + \frac{3a}{a+b+c} = 0,$$

and we deduce that $a = b = c$.

Problem 2.51 Let ABC be an acute triangle. The interior bisectors of the angles $\angle B$ and $\angle C$ meet the opposite sides in the points L and M, respectively. Prove that there exists a point K in the interior of the side BC such that triangle KLM is equilateral if and only if $\angle A = 60°$.

Solution Let us first notice that if M is a point on the bisector of the angle XOY and $P \in OX$, $Q \in OY$ such that $MP = MQ$, then either $\angle OPM = \angle OQM$, or $\angle OPM + \angle OQM = 180°$ (see Fig. 5.38).

Returning to the problem, let us suppose such a point K exists (Fig. 5.39). Then M lies on the bisector of $\angle ACB$ and $ML = MK$, hence $\angle ALM = \angle BKM$ or $\angle ALM + \angle BKM = 180°$. But in the last case, it follows that the quadrilateral $MKCL$ is cyclic, so $\angle C = 180° - \angle KML = 120°$, which is a contradiction since triangle ABC is acute.

It follows that $\angle ALM = \angle BKM$, and, in a similar way, $\angle LKC = \angle LMA$. In triangle AML, we have

$$\angle AML + \angle ALM + \angle A = 180°.$$

Fig. 5.39

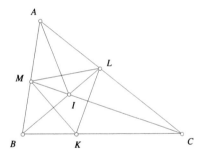

On the other hand,

$$\angle BKM + \angle LKC + \angle MKL = 180°.$$

We derive that $\angle A = \angle MKL = 60°$.

Conversely, suppose $\angle A = 60°$. Let $K \in (BC)$ such that $MK \perp BL$. Because BL is the bisector of $\angle B$, it follows that BL is the perpendicular bisector of MK, so $LM = LK$.

Denote by I the intersection point of BL and CM. Clearly, AI bisects $\angle A$ and a short computation shows that $\angle MIL = 120°$, hence the quadrilateral $AMIL$ is cyclic. Then $\angle MLI = \angle MAI = 30°$, thus we have $\angle KML = 60°$. It follows that KLM is an isosceles triangle with a $60°$ angle, hence it is equilateral.

Problem 2.52 Let $P_1 P_2 \ldots P_n$ be a convex polygon with the following property: for any two vertices P_i and P_j, there exists a vertex P_k such that $\angle P_i P_k P_j = 60°$. Prove that the polygon is an equilateral triangle.

Solution Consider the vertices P_i and P_j such that the segment $P_i P_j$ has minimal length and P_k such that $\angle P_i P_k P_j = 60°$. It follows that triangle $P_i P_k P_j$ is equilateral. Denote it by ABC. Similarly, let $A'B'C'$ the equilateral triangle with sides of maximal length. Because the polygon is convex, the points A', B', C' must lie in the set $D_A \cup D_B \cup D_C$ (see Fig. 5.40).

After a short analysis, we conclude that the points A', B' and C' coincide with A, B, C, and it follows that the polygon is an equilateral triangle.

Problem 2.53 From a point on the circumcircle of an equilateral triangle ABC parallels to the sides BC, CA and AB are drawn, intersecting the sides CA, AB and BC at the points M, N, P, respectively. Prove that the points M, N, P are collinear.

Solution It suffices to prove that $\angle CPM = \angle BPN$ (see Fig. 5.41).

Observe that the quadrilateral $CMDP$ is cyclic. Indeed, since DP is parallel to AB, the angle $\angle DPC$ equals $120°$, hence

$$\angle DPC + \angle DMC = 120° + 60° = 180°.$$

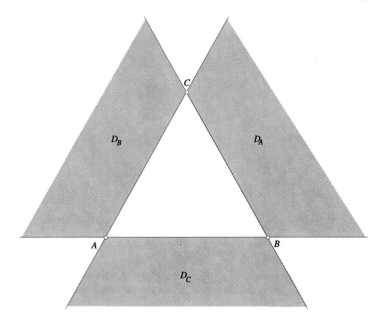

Fig. 5.40

Fig. 5.41

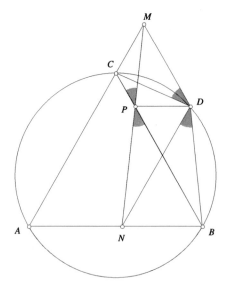

Consequently, $\angle CPM = \angle CDM$. In a similar way, we obtain $\angle BPN = \angle BDN$. But $\angle BDC = 120°$ ($ABDC$ is cyclic) and $\angle MDN = 120°$ ($AMDN$ is an isosceles trapezoid). It follows that $\angle CDM = \angle BDN$, hence $\angle CPM = \angle BPN$.

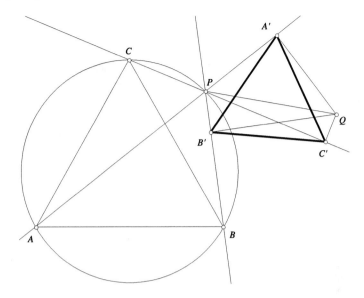

Fig. 5.42

Problem 2.54 Let P be a point on the circumcircle of an equilateral triangle ABC. Prove that the projections of any point Q on the lines PA, PB and PC are the vertices of an equilateral triangle.

Solution Let A', B', C' be the projections of a point Q on the lines PA, PB, PC, respectively. If the points are located as in Fig. 5.42, then $\angle A'PB' = 120°$, $\angle A'PC' = \angle B'PC' = 60°$.

From the construction it follows that the quadrilaterals $A'PB'Q, A'PC'Q$ and $B'PC'Q$ are cyclic, so $\angle A'PQ = \angle A'B'Q, \angle A'PQ = \angle A'C'Q$. Then $\angle A'B'Q = \angle A'C'Q$ and this shows that $A'B'C'Q$ is also cyclic. It results that

$$\angle A'B'C' = 180° - \angle A'QC' = \angle A'PC' = 60°,$$
$$\angle A'C'B' = \angle A'QB' = 180 - \angle A'PB' = 60°,$$

hence triangle $A'B'C'$ is equilateral.

5.6 The "Carpets" Theorem

Problem 2.56 Let $ABCD$ be a parallelogram. The points M, N and P are chosen on the segments BD, BC and CD, respectively, so that $CNMP$ is a parallelogram. Let $E = AN \cap BD$ and $F = AP \cap BD$. Prove that $[AEF] = [DFP] + [BEN]$.

Fig. 5.43

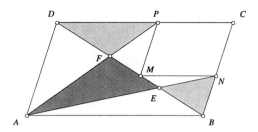

Solution The carpets are the triangle ABD and the quadrilateral $ANCP$. We have

$$[ABD] = \frac{1}{2}[ABCD]$$

and

$$[ANCP] = [ANC] + [ACP].$$

Observe that

$$\frac{[ANC]}{[ABCD]} = \frac{[ANC]}{2[ABC]} = \frac{NC}{2BC} = \frac{MD}{2BD}$$

and

$$\frac{[ACP]}{[ABCD]} = \frac{[ACP]}{2[ACD]} = \frac{CP}{2CD} = \frac{MB}{2BD}.$$

Adding up, we obtain

$$\frac{[ANC] + [ACP]}{[ABCD]} = \frac{MD + MB}{2BD} = \frac{1}{2}.$$

It follows that

$$[ABD] + [ANCP] = [ABCD],$$

so the area of the common part of the carpets, that is, $[AEF]$ equals the area of $ABCD$ remained uncovered and this is $[DFP] + [BEN]$ (Fig. 5.43).

Observation Another proof uses a well known property of a trapezoid. Because $[ADP] = [ADM]$ it follows that $[DFP] = [AFM]$. Similarly, $[BEN] = [AEM]$ and the result follows by adding these equalities.

Problem 2.57 Consider the quadrilateral $ABCD$. The points M, N, P and Q are the midpoints of the sides AB, BC, CD and DA. Let $X = AP \cap BQ$, $Y = BQ \cap CM$, $Z = CM \cap DN$ and $T = DN \cap AP$. Prove that

$$[XYZT] = [AQX] + [BMY] + [CNZ] + [DPT].$$

Fig. 5.44

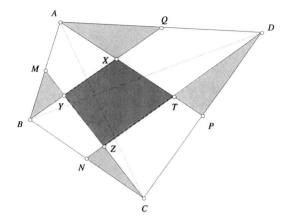

Solution The carpets are the quadrilaterals $AMCP$ and $BNDQ$. Because their common part is $XYZT$ all we have to prove is that

$$[AMCP] + [BNDQ] = [ABCD].$$

We have

$$[AMCP] = [AMC] + [ACP] = \frac{1}{2}[ABC] + \frac{1}{2}[ACD] = \frac{1}{2}[ABCD].$$

Similarly,

$$[BNDQ] = \frac{1}{2}[ABCD],$$

hence

$$[AMCP] + [BNDQ] = [ABCD],$$

as desired (Fig. 5.44).

Problem 2.58 Through the vertices of the smaller base AB of the trapezoid $ABCD$ two parallel lines are drawn, intersecting the segment CD. These lines and the trapezoid's diagonals divide it into 7 triangles and a pentagon. Show that the area of the pentagon equals the sum of areas of the three triangles sharing a common side with the trapezoid.

Solution Let A' and B' be the intersections of the parallel lines with CD. Take as carpets triangles $AA'C$ and $BB'D$. Their common part is the pentagon and the uncovered part of the trapezoid is the union of the three triangles sharing a common side with the trapezoid (Fig. 5.45). We then have to prove that

$$\left[AA'C\right] + \left[BB'D\right] = [ABCD].$$

Fig. 5.45

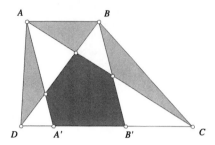

Fig. 5.46

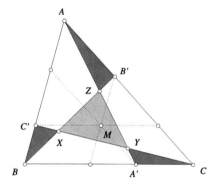

If we denote the trapezoid's altitude by h, we have $[AA'C] = \frac{1}{2}h \cdot A'C$ and $[BB'D] = \frac{1}{2}h \cdot B'D$, thus

$$[AA'C] + [BB'D] = \frac{1}{2}h \cdot (A'C + B'D) = \frac{1}{2}h \cdot (A'B' + B'C + A'B' + A'D)$$

$$= \frac{1}{2}h \cdot (CD + A'B') = \frac{1}{2}h \cdot (CD + AB) = [ABCD].$$

Problem 2.59 Let M be a point in the interior of triangle ABC. Three lines are drawn through M, parallel to triangle's sides, determining three trapezoids. One draws a diagonal in each trapezoid such that they have no common endpoints, dividing thus ABC into seven parts, four of them being triangles. Prove that the area of one of the four triangles equals the sum of the areas of the other three.

Solution We arrange three carpets. With the notations in Fig. 5.46, observe that if the carpets are ABB', BCC' and CAA', then triangles $AB'Z$, $BC'X$ and $CA'Y$ are covered twice, while XYZ remains uncovered.

Thus, the equality

$$[AB'Z] + [BC'X] + [CA'Y] = [XYZ]$$

Fig. 5.47

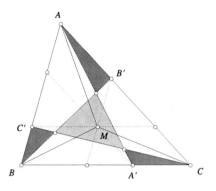

holds if and only if the sum of the carpets' areas equals the area of ABC. But this is very simple to prove if we notice that (Fig. 5.47)

$$[ABB'] = [ABM],$$
$$[BCC'] = [BCM]$$

and

$$[CAA'] = [CAM].$$

We obviously have

$$[ABM] + [BCM] + [CAM] = [ABC].$$

5.7 Quadrilaterals with an Inscribed Circle

Problem 2.60 Prove that if in the quadrilateral $ABCD$ is inscribed a circle with center O, then the sum of the angles $\angle AOB$ and $\angle COD$ equals $180°$ (Fig. 5.48).

Solution We know that the quadrilateral's angles bisectors intersect at O. Thus

$$\angle AOB = 180° - \angle ABO - \angle BAO = 180° - \frac{\angle A + \angle B}{2}.$$

Similarly,

$$\angle COD = 180° - \frac{\angle C + \angle D}{2}.$$

Adding these equalities, we obtain

$$\angle AOB + \angle COD = 360° - \frac{\angle A + \angle B + \angle C + \angle D}{2} = 360° - \frac{360°}{2} = 180°.$$

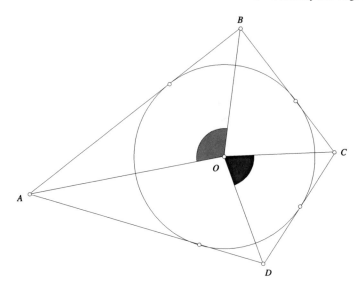

Fig. 5.48

Fig. 5.49

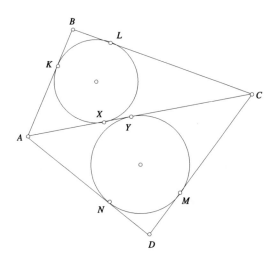

Problem 2.61 Let $ABCD$ be a quadrilateral with an inscribed circle. Prove that the circles inscribed in triangles ABC and ADC are tangent to each other.

Solution Suppose that the circles inscribed in triangles ABC and ADC touch AC at the points X and Y, respectively. We have to prove that $X = Y$. With the notation in Fig. 5.49, we have $AX = AK$ (the tangents to a circle drawn from a point are equal), $BK = BL$, $CL = CX$, $CY = CM$, $DM = DN$, $AN = AY$.

Because $ABCD$ has an inscribed circle, $AB + CD = AD + BC$, so

$$AK + BK + CM + MD = AN + DN + BL + CL.$$

Fig. 5.50

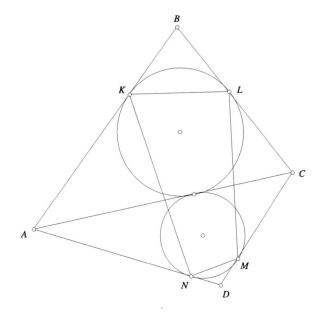

Using the previous equalities, we obtain

$$AX + CY = AY + CX.$$

Adding the obvious equality

$$AX + XC = AY + YC$$

yields $2AX = 2AY$, hence $X = Y$.

Observation It can be proven that the points K, L, M, N are the vertices of a cyclic quadrilateral (Fig. 5.50).

Indeed, if we draw the circle inscribed in $ABCD$ and denote by K', L', M' and N' the tangency points with the quadrilateral's sides, it is not difficult to see that the sides of $KLMN$ and $K'L'M'N'$ are parallel, hence their corresponding angles are equal. But $K'L'M'N'$ is cyclic, thus its opposite angles add up to 180°. The conclusion follows (Fig. 5.51).

Problem 2.62 Let $ABCD$ be a convex quadrilateral. Suppose that the lines AB and CD intersect at E and the lines AD and BC intersect at F, such that the points E and F lie on opposite sides of the line AC. Prove that the following statements are equivalent:

(i) a circle is inscribed in $ABCD$;
(ii) $BE + BF = DE + DF$;
(iii) $AE - AF = CE - CF$.

Fig. 5.51

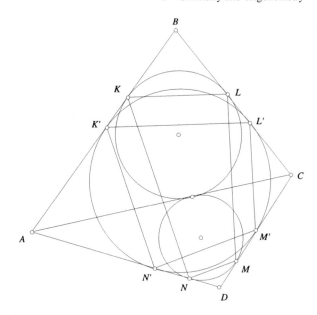

Solution Suppose a circle is inscribed in the quadrilateral $ABCD$ and touches its sides at the points K, L, M, N (Fig. 5.52).

Observe that

$$BE + BF = EK - BK + BL + LF = EM + NF$$

$$= EM + ND + DF = EM + MD + DF = DE + DF.$$

We used again the fact that the tangents to a circle from a point are equal, so $BK = BL$, $EK = EM$, $LF = NF$ and $ND = MD$.

In a similar way we have

$$AE - AF = AK + EK - AN - NF = EK - NF$$

$$= EM - LF = CE + CM - CL - CF = CE - CF.$$

Conversely, if, for instance, $BE + BF = DE + DF$, draw the circle tangent to AB, BC, and AF. If this circle is not tangent to CD as well, draw from E a tangent to the circle which intersects AF at D'. Then $BE + BF = D'E + D'F$, and we deduce $D = D'$, a contradiction. We conclude that $ABCD$ has an inscribed circle.

Problem 2.63 Let $ABCD$ be a convex quadrilateral. Suppose that the lines AB and CD intersect at E and the lines AD and BC intersect at F. Let M and N be two arbitrary points on the line segments AB and BC, respectively. The line EN intersects AF and MF at P and R. The line MF intersects CE at Q. Prove that if the quadrilaterals $AMRP$ and $CNRQ$ have inscribed circles, then $ABCD$ has an inscribed circle.

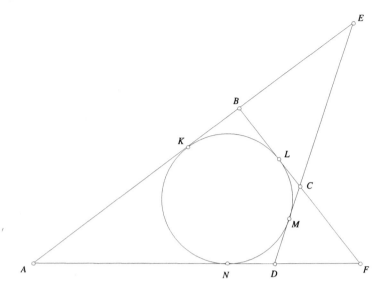

Fig. 5.52

Fig. 5.53

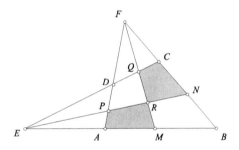

Solution Suppose the points are located as in Fig. 5.53, the other cases being similar. Because $AMRP$ has an inscribed circle, it follows from the preceding problem that $AE + AF = RE + RF$. Analogously, since $CNRQ$ has an inscribed circle, we have $RE + RF = CE + CF$. We obtain $AE + AF = CE + CF$ and this implies that $ABCD$ has an inscribed circle.

Problem 2.64 The points A_1, A_2, C_1 and C_2 are chosen in the interior of the sides CD, BC, AB and AD of the convex quadrilateral $ABCD$. Denote by M the point of intersection of the lines AA_2 and CC_1 and by N the point of intersection of the lines AA_1 and CC_2. Prove that if one can inscribe circles in three of the four quadrilaterals $ABCD$, A_2BC_1M, $AMCN$ and A_1NC_2D, then a circle can be also inscribed in the fourth one.

Solution Let $\alpha = AB - BC - AM + CM$. From the previous problem it follows that a circle can be inscribed in the quadrilateral A_2BC_1M if and only if $\alpha = 0$.

Fig. 5.54

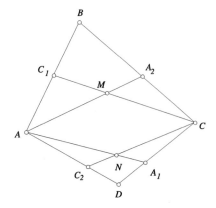

Analogously, if we set

$$\beta = CD - AD - CN + AN,$$

then $\beta = 0$ if and only if A_1DC_2N has an inscribed circle. Setting

$$\gamma = AM - CM + CN - AN,$$

$$\delta = BC - AB + AD - CD,$$

from the theorem of Pithot it follows that $\gamma = 0$ and $\delta = 0$ are necessary and sufficient conditions for the existence of an inscribed circle in $AMCN$ and $ABCD$, respectively.

Now, simply observe that $\alpha + \beta + \gamma + \delta = 0$, so that if three of the four numbers are zero, then so is also the fourth one. It follows that if one can inscribe circles in three of the four quadrilaterals, a circle can also be inscribed in the fourth one (Fig. 5.54).

Problem 2.65 A line cuts a quadrilateral with an inscribed circle into two polygons with equal areas and equal perimeters. Prove that the line passes through the center of the inscribed circle.

Solution The line cuts the quadrilateral either into two quadrilaterals, or into a triangle and a pentagon. The reasoning is basically the same in both cases, so we assume that the line intersect the sides AB and CD at the points X and Y (Fig. 5.55). Let O be the center of the inscribed circle. Because $AXYD$ and $BXYC$ have the same perimeter, it follows that

$$AX + AD + DY = BX + BC + CY.$$

Multiplying this equality with $\frac{1}{2}R$ (R being the radius of the inscribed circle) yields

$$[OAX] + [OAD] + [ODY] = [OBX] + [OBC] + [OCY],$$

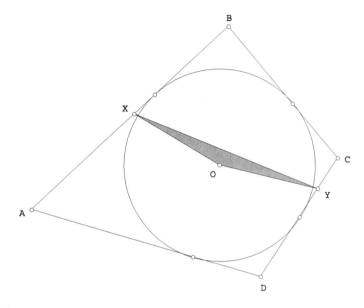

Fig. 5.55

that is,

$$[OXADY] = [OXBCY].$$

Because

$$[OXADY] + [OXBCY] = [ABCD],$$

it follows that

$$[OXADY] = [OXBCY] = \frac{1}{2}[ABCD].$$

Now, suppose by way of contradiction that the line XY does not pass through O. Suppose O lies in the interior of $AXYD$ (the case in which O is in the interior of $BXYC$ is similar). We have

$$[BXYC] = \frac{1}{2}[ABCD] = [BXOYC],$$

but $[BXOYC] = [BXYC] + [OXY]$, hence $[OXY] = 0$, which is a contradiction.

Problem 2.66 In the convex quadrilateral $ABCD$ we have $\angle B = \angle C = 120°$, and

$$AB^2 + BC^2 + CD^2 = AD^2.$$

Prove that $ABCD$ has an inscribed circle.

Solution Suppose AB and CD intersect at E. Since $\angle B = \angle C = 120°$, the triangle BCE is equilateral. Denote by x its side length. Applying the law of cosines in triangle ADE yields

$$(AB + x)^2 + (CD + x)^2 - (AB + x)(CD + x) = AD^2.$$

Expanding and using the equality from the hypothesis lead to

$$AB \cdot x + CD \cdot x - AB \cdot CD = 0.$$

But then

$$\begin{aligned}(AB + CD - x)^2 &= AB^2 + CD^2 + x^2 - 2(AB \cdot x + CD \cdot x - AB \cdot CD) \\ &= AB^2 + CD^2 + x^2 \\ &= AD^2.\end{aligned}$$

It follows that

$$AB + CD = AD + x = AD + BC,$$

therefore $ABCD$ has an inscribed circle.

Problem 2.67 Let $ABCD$ be a quadrilateral circumscribed about a circle, whose interior and exterior angles are at least $60°$. Prove that

$$\frac{1}{3}|AB^3 - AD^3| \le |BC^3 - CD^3| \le 3|AB^3 - AD^3|.$$

When does equality hold?

Solution By symmetry, it suffices to prove the first inequality.

Since $ABCD$ has an inscribed circle, we have $AB + CD = AD + BC$, or, equivalently, $AB - AD = BC - CD$. Therefore, the inequality we want to prove is equivalent to

$$\frac{1}{3}(AB^2 + AB \cdot AD + AD^2) \le BC^2 + BC \cdot CD + CD^2.$$

From the hypothesis we have $60° \le \angle A, \angle C \le 120°$, therefore $\frac{1}{2} \ge \cos A$, $\cos C \ge -\frac{1}{2}$. Applying the law of cosines in triangle ABD yields

$$BD^2 = AB^2 - 2AB \cdot AD \cos A + AD^2 \ge AB^2 - AB \cdot AD + AD^2.$$

But

$$AB^2 - AB \cdot AD + AD^2 \ge \frac{1}{3}(AB^2 + AB \cdot AD + AD^2), \tag{5.3}$$

the latter being equivalent to

$$(AB - AD)^2 \ge 0.$$

Applying again the law of cosines in triangle BCD yields

$$BD^2 = BC^2 - 2BC \cdot CD \cos C + CD^2 \le BC^2 + BC \cdot CD + CD^2. \tag{5.4}$$

Combining (5.3) and (5.4) gives the desired result.

The equality holds if and only if $AB = AD$, which also implies $BC = CD$ (that is, the given quadrilateral is a kite).

5.8 Dr. Trig Learns Complex Numbers

Problem 2.71 Let a, b, c be real numbers such that

$$\cos a + \cos b + \cos c = \sin a + \sin b + \sin c = 0.$$

Prove that

$$\cos(a + b + c) = \frac{1}{3}(\cos 3a + \cos 3b + \cos 3c),$$

$$\sin(a + b + c) = \frac{1}{3}(\sin 3a + \sin 3b + \sin 3c).$$

Solution Let $x = \cos a + i \sin a$, $y = \cos b + i \sin b$ and $z = \cos c + i \sin c$. We derive from the problem statement that $x + y + z = 0$. We use the identity

$$x^3 + y^3 + z^3 - 3xyz = (x + y + z)(x^2 + y^2 + z^2 - xy - xz - yz),$$

which is valid for any numbers x, y and z (see the Algebra part of this book, Sect. 1.1). It follows that

$$x^3 + y^3 + z^3 - 3xyz = 0,$$

hence

$$xyz = \frac{1}{3}(x^3 + y^3 + z^3).$$

But this means that

$$\cos(a + b + c) + i \sin(a + b + c) = \frac{1}{3}\sum(\cos 3a + i \sin 3a),$$

and we obtain the requested equalities by identifying the real parts and the non-real parts in the above equality.

Problem 2.72 Find the value of the product $\cos 20° \cos 40° \cos 80°$.

Solution Let $z = \cos 20° + i \sin 20°$. Then

$$\cos 20° = \frac{1}{2}\left(z + \frac{1}{z}\right), \qquad \cos 40° = \frac{1}{2}\left(z^2 + \frac{1}{z^2}\right), \qquad \cos 80° = \frac{1}{2}\left(z^4 + \frac{1}{z^4}\right).$$

Also, $z^9 = \cos 180° + i \sin 180° = -1$, thus

$$\frac{1}{z} = -z^8, \qquad \frac{1}{z^2} = -z^7, \qquad \frac{1}{z^4} = -z^5.$$

The product equals

$$\frac{1}{8}(z - z^8)(z^2 - z^7)(z^4 - z^5)$$

$$= \frac{1}{8}(z^3 - z^8 - z^{10} + z^{15})(z^4 - z^5)$$

$$= \frac{1}{8}(z^3 - z^8 + z - z^6)(z^4 - z^5)$$

$$= \frac{1}{8}(z^7 - z^8 - z^{12} + z^{13} + z^5 - z^6 - z^{10} + z^{11})$$

$$= \frac{1}{8}(z - z^2 + z^3 - z^4 + z^5 - z^6 + z^7 - z^8)$$

$$= \frac{1}{8}\left(1 - \frac{z^9 + 1}{z + 1}\right) = \frac{1}{8}.$$

Problem 2.73 Prove that

$$\frac{1}{\cos 6°} + \frac{1}{\sin 24°} + \frac{1}{\sin 48°} = \frac{1}{\sin 12°}.$$

Solution If we denote $z = \cos 6° + i \sin 6°$, then $z^{15} = \cos 90° + i \sin 90° = i$. We have

$$\cos 6° = \frac{z^2 + 1}{2z}, \qquad \sin 12° = \frac{z^4 - 1}{2i z^2}, \qquad \sin 24° = \frac{z^8 - 1}{2i z^4},$$

$$\sin 48° = \frac{z^{16} - 1}{2i z^8}.$$

The equality to be proved becomes

$$\frac{2z}{z^2 + 1} - \frac{2i z^2}{z^4 - 1} + \frac{2i z^4}{z^8 - 1} + \frac{2i z^8}{z^{16} - 1} = 0.$$

Multiplying by $z^{16} - 1$, we obtain, after a short computation

$$z^{16} - 1 - iz(z^{14} + 1) = 0 \quad \Longleftrightarrow \quad iz - 1 - i^2 - iz = 0,$$

which is obvious.

Problem 2.74 Prove that

$$\cos \frac{2\pi}{7} + \cos \frac{4\pi}{7} + \cos \frac{6\pi}{7} + \frac{1}{2} = 0.$$

Solution If $z = \cos \frac{2\pi}{7} + i \sin \frac{2\pi}{7}$, then $z^7 = 1$. The equality to be proved becomes

$$\frac{1}{2}\left(z + \frac{1}{z}\right) + \frac{1}{2}\left(z^2 + \frac{1}{z^2}\right) + \frac{1}{2}\left(z^3 + \frac{1}{z^3}\right) + \frac{1}{2} = 0.$$

Multiplying by $2z^3$ and rearranging the terms, we get

$$z^6 + z^5 + z^4 + z^3 + z^2 + z + 1 = 0,$$

or

$$\frac{z^7 - 1}{z - 1} = 0,$$

which is obvious.

Problem 2.75 Prove the equality

$$\sin\frac{\pi}{2n} \sin\frac{2\pi}{2n} \cdots \sin\frac{(n-1)\pi}{2n} = \frac{\sqrt{n}}{2^{n-1}}.$$

Solution Consider the polynomial $P(X) = X^{2n} - 1$. Its roots are the numbers $x_k = \cos\frac{k\pi}{n} + i\sin\frac{k\pi}{n}$, with $k = 0, 1, \ldots, 2n - 1$. It follows that $P(X) = (X - x_0) \times (X - x_1)\cdots(X - x_{2n-1})$. We observe that, except $x_0 = 1$ and $x_n = -1$, all the other roots are non-real complex numbers and that, for $1 \le k \le n - 1$, $\bar{x}_k = x_{2n-k}$. Thus, we can write

$$P(X) = (X^2 - 1)\prod_{k=1}^{n-1}(X - x_k)(X - \bar{x}_k)$$

$$= (X^2 - 1)\prod_{k=1}^{n-1}\left(x^2 - (x_k + \bar{x}_k)x + x_k\bar{x}_k\right).$$

Now, $x_k + \bar{x}_k = 2\cos\frac{k\pi}{n}$ and $x_k\bar{x}_k = 1$. Dividing both sides by $x^2 - 1$, we obtain

$$x^{2n-2} + x^{2n-4} + \cdots + x^4 + x^2 + 1 = \prod_{k=1}^{n-1}\left(x^2 - 2x\cos\frac{k\pi}{n} + 1\right).$$

For $x = 1$, the above equality becomes

$$n = 2^{n-1}\prod_{k=1}^{n-1}\left(1 - \cos\frac{k\pi}{n}\right) = 2^{n-1}\prod_{k=1}^{n-1}2\sin^2\frac{k\pi}{2n}$$

$$= \left(2^{n-1}\prod_{k=1}^{n-1}\sin\frac{k\pi}{2n}\right)^2.$$

Since $\sin\frac{k\pi}{2n} > 0$ for $k = 1, \ldots, n - 1$, we obtain

$$\prod_{k=1}^{n-1}\sin\frac{k\pi}{2n} = \frac{\sqrt{n}}{2^{n-1}}.$$

Problem 2.76 Solve the equation

$$\sin x + \sin 2x + \sin 3x = \cos x + \cos 2x + \cos 3x.$$

Solution If $z = \cos x + i \sin x$, the equation becomes

$$\frac{z^2 - 1}{2iz} + \frac{z^4 - 1}{2iz^2} + \frac{z^6 - 1}{2iz^3} = \frac{z^2 + 1}{2z} + \frac{z^4 + 1}{2z^2} + \frac{z^6 + 1}{2z^3}.$$

Multiplying by $2iz^3$ and grouping the terms, we obtain

$$z^6 + z^5 + z^4 - i\left(z^2 + z + 1\right) = 0,$$

or

$$\left(z^4 - i\right)\left(z^2 + z + 1\right) = 0.$$

If $z^2 + z + 1 = 0$, then $z = \frac{-1 \pm i\sqrt{3}}{2} = \cos\frac{\pm 2\pi}{3} + i \sin\frac{\pm 2\pi}{3}$, hence $x = \frac{\pm 2\pi}{3} + 2k\pi$, with $k \in \mathbf{Z}$. If $z^4 = i$, then $\cos 4x + i \sin 4x = \cos\frac{\pi}{2} + i \sin\frac{\pi}{2}$ and it follows that $4x = \frac{\pi}{2} + 2k\pi$, hence $x = \frac{\pi}{8} + \frac{k\pi}{2}$, with $k \in \mathbf{Z}$.

Problem 2.77 Prove that

$$\cos\frac{\pi}{5} = \frac{1 + \sqrt{5}}{4}.$$

Solution If we denote by $z = \cos\frac{\pi}{5} + i \sin\frac{\pi}{5}$, then $z^5 = -1$, which is equivalent to

$$(z + 1)\left(z^4 - z^3 + z^2 - z + 1\right) = 0.$$

Because $z \neq -1$, we obtain $z^4 - z^3 + z^2 - z + 1 = 0$. After dividing by $z^2 \neq 0$, it follows that

$$z^2 + \frac{1}{z^2} - \left(z + \frac{1}{z}\right) + 1 = 0.$$

If $x = \cos\frac{\pi}{5}$, then

$$z + \frac{1}{z} = 2x, \qquad z^2 + \frac{1}{z^2} = \left(z + \frac{1}{z}\right)^2 - 2 = 4x^2 - 2.$$

It follows that x is a (positive) root of the equation

$$4x^2 - 2x - 1 = 0,$$

hence

$$x = \frac{1 + \sqrt{5}}{4}.$$

Chapter 6
Number Theory and Combinatorics

6.1 Arrays of Numbers

Problem 3.4 Prove that the sum of any n entries of the table

1	$\frac{1}{2}$	$\frac{1}{3}$	$\cdots$	$\frac{1}{n}$
$\frac{1}{2}$	$\frac{1}{3}$	$\frac{1}{4}$	$\cdots$	$\frac{1}{n+1}$
$\vdots$				
$\frac{1}{n}$	$\frac{1}{n+1}$	$\frac{1}{n+2}$	$\cdots$	$\frac{1}{2n-1}$

situated in different rows and different columns is not less than 1.

Solution Denoting by a_{ij} the entry in the ith row and jth column of the array, we have

$$a_{ij} = \frac{1}{i+j-1},$$

for all $i, j, 1 \leq i, j \leq n$. Choose n entries situated in different rows and different columns. It follows that from each row and each column exactly one number is chosen. Let $a_{1j_1}, a_{2j_2}, \ldots, a_{nj_n}$ be the chosen numbers, where $j_1, j_2, \ldots, j_n$ is a permutation of indices $1, 2, \ldots, n$. We have

$$\sum_{k=1}^{n} \frac{1}{a_{kj_k}} = \sum_{k=1}^{n}(k + j_k - 1) = \sum_{k=1}^{n} k + \sum_{k=1}^{n} j_k - n.$$

But

$$\sum_{k=1}^{n} j_k = \sum_{k=1}^{n} k = \frac{n(n+1)}{2}$$

T. Andreescu, B. Enescu, *Mathematical Olympiad Treasures*,
DOI 10.1007/978-0-8176-8253-8_6, © Springer Science+Business Media, LLC 2011

since $j_1, j_2, \ldots, j_n$ is a permutation of indices $1, 2, \ldots, n$. It follows that

$$\sum_{k=1}^{n} \frac{1}{a_{kj_k}} = n^2.$$

The Cauchy–Schwarz inequality yields

$$(x_1 + x_2 + \cdots + x_n)\left(\frac{1}{x_1} + \frac{1}{x_2} + \cdots + \frac{1}{x_n}\right) \geq n^2,$$

for all positive real numbers $x_1, x_2, \ldots, x_n$. Taking $x_k = a_{kj_k}$ and using the above equality, we obtain

$$\sum_{k=1}^{n} a_{kj_k} \geq 1$$

as desired.

Problem 3.5 The entries of an $n \times n$ array of numbers are denoted by $a_{ij}, 1 \leq i, j \leq n$. The sum of any n entries situated on different rows and different columns is the same. Prove that there exist numbers $x_1, x_2, \ldots, x_n$ and $y_1, y_2, \ldots, y_n$, such that

$$a_{ij} = x_i + y_j,$$

for all i, j.

Solution Consider n entries situated on different rows and different columns $a_{ij_i}, i = 1, 2, \ldots, n$. Fix k and $l, 1 \leq k < l \leq n$ and replace a_{kj_k} and a_{lj_l} with a_{kj_l} and a_{lj_k}, respectively. It is not difficult to see that the new n entries are still situated on different rows and different columns. Because the sums of the two sets of n entries are equal, it follows that

$$a_{kj_k} + a_{lj_l} = a_{kj_l} + a_{lj_k}. \tag{$*$}$$

Now, denote $x_1, x_2, \ldots, x_n$ the entries in the first column of the array and by $x_1, x_1 + y_2, x_1 + y_3, \ldots, x_1 + y_n$ the entries in the first row (in fact, we have defined $x_k = a_{k1}$, for all k, $y_0 = 0$ and $y_k = a_{1k} - a_{k1}$ for all $k \geq 2$).

x_1	$x_1 + y_2$	$\cdots$	$x_1 + y_j$	$\cdots$	$x_1 + y_n$
x_2					
$\vdots$					
x_i			a_{ij}		
$\vdots$					
x_n					

The equality

$$a_{ij} = x_i + y_j,$$

stands for all i, j with $i = 1$ or $j = 1$. Now, consider $i, j > 1$. From $(*)$ we deduce

$$a_{11} + a_{ij} = a_{1j} + a_{i1}$$

hence

$$x_1 + a_{ij} = x_i + x_1 + y_j$$

or

$$a_{ij} = x_i + y_j,$$

for all i, j, as desired.

Problem 3.6 In an $n \times n$ array of numbers all rows are different (two rows are different if they differ in at least one entry). Prove that there is a column which can be deleted in such a way that the remaining rows are still different.

Solution We prove by induction on k the following statement: at least $n - k + 1$ columns can be deleted in such a way that the first k rows are still different. For $k = 2$ the assertion is true. Indeed, the first two rows differ in at least one place, so we can delete the remaining $n - 1$ columns. Suppose the assertion is true for k, that is we can delete $n - k + 1$ columns and the first k rows are still different. If after the deletion of the columns the $(k + 1)$th row is different from all first k rows, we can put back any of the deleted columns and remain with $n - k$ deleted columns and $k + 1$ different rows. If after the deletion the $(k + 1)$th row coincides with one of the first rows, then we put back the column in which the two rows differ in the original array. For $k = n$ we obtain the desired result.

Problem 3.7 The positive integers from 1 to n^2 $(n \geq 2)$ are placed arbitrarily on squares of an $n \times n$ chessboard. Prove that there exist two adjacent squares (having a common vertex or a common side) such that the difference of the numbers placed on them is not less than $n + 1$.

Solution Suppose the contrary: the difference of the numbers placed in any adjacent squares is less than $n + 1$. If we place a king on a square of the chessboard, it can reach any other square in at most $n - 1$ moves through adjacent squares. Place a king in the square with number 1 and move it to the square with number n^2 in at most $n - 1$ moves. At each move, the difference between the numbers in the adjacent squares is less than $n + 1$, hence the difference between n^2 and 1 is less than $(n + 1)(n - 1) = n^2 - 1$, a contradiction.

Problem 3.8 A positive integer is written in each square of an $n^2 \times n^2$ chess board. The difference between the numbers in any two adjacent squares (sharing an edge) is less than or equal to n. Prove that at least $\lfloor \frac{n}{2} \rfloor + 1$ squares contain the same number.

Solution Consider the smallest and largest numbers a and b on the board. They are separated by at most $n^2 - 1$ squares horizontally and $n^2 - 1$ vertically, so there is a path from one to the other with length at most $2(n^2 - 1)$. Then since any two successive squares differ by at most n, we have $b - a \leq 2(n^2 - 1)n$. But since all numbers on the board are integers lying between a and b, only $2(n^2 - 1)n + 1$ distinct numbers can exist; and because

$$n^4 > \left(2(n^2 - 1)n + 1\right)\frac{n}{2},$$

more than $\frac{n}{2}$ squares contain the same number, as needed.

Problem 3.9 The numbers $1, 2, \ldots, 100$ are arranged in the squares of an 10×10 table in the following way: the numbers $1, \ldots, 10$ are in the bottom row in increasing order, numbers $11, \ldots, 20$ are in the next row in increasing order, and so on. One can choose any number and two of its neighbors in two opposite directions (horizontal, vertical, or diagonal). Then either the number is increased by 2 and its neighbors are decreased by 1, or the number is decreased by 2 and its neighbors are increased by 1. After several such operations the table again contains all the numbers $1, 2, \ldots, 100$. Prove that they are in the original order.

Solution Label the table entry in the ith row and jth column by a_{ij}, where the bottom-left corner is in the first row and first column. Let $b_{ij} = 10(i-1) + j$ be the number originally in the ith row and jth column.

Observe that

$$P = \sum_{i,j=1}^{10} a_{ij}b_{ij}$$

is invariant. Indeed, every time entries a_{mn}, a_{pq}, a_{rs} are changed (with $m + r = 2p$ and $n + s = 2q$), P increases or decreases by $b_{mn} - 2b_{pq} + b_{rs}$, but this equals

$$10\left((m-1) + (r-1) - 2(p-1)\right) + (n + s - 2q) = 0.$$

In the beginning,

$$P = \sum_{i,j=1}^{10} b_{ij}b_{ij}$$

at the end, the entries a_{ij} equal the b_{ij} in some order, and we now have

$$P = \sum_{i,j=1}^{10} a_{ij}b_{ij}$$

By the rearrangement inequality, this is at least $P = \sum_{i,j=1}^{10} a_{ij}b_{ij}$ with equality only when each $a_{ij} = b_{ij}$. The equality does occur since P is invariant. Therefore the a_{ij} do indeed equal the b_{ij} in the same order, and thus the entries $1, 2, \ldots, 100$ appear in their original order.

Problem 3.10 Prove that one cannot arrange the numbers from 1 to 81 in a 9×9 table such that for each i, $1 \leq i \leq 9$ the product of the numbers in row i equals the product of the numbers in column i.

Solution The key observation is the following: if row k contains a prime number $p > 40$, then the same number must be contained by column k, as well. Therefore, all prime numbers from 1 to 81 must lie on the main diagonal of the table. However, this is impossible, since there are 10 such prime numbers: $41, 43, 47, 53, 59, 61, 67, 71, 73$, and 79.

Problem 3.11 The entries of a matrix are integers. Adding an integer to all entries on a row or on a column is called an operation. It is given that for infinitely many integers N one can obtain, after a finite number of operations, a table with all entries divisible by N. Prove that one can obtain, after a finite number of operations, the zero matrix.

Solution Suppose the matrix has m rows and n columns and its entries are denoted by a_{hk}. Fix some j, $1 < j \leq n$, and consider an arbitrary i, $1 < i \leq m$. The expression

$$E_{ij} = a_{11} + a_{ij} - a_{i1} - a_{1j}$$

is an invariant for our operation. Indeed, adding k to the first row, it becomes

$$(a_{11} + k) + a_{ij} - a_{i1} - (a_{1j} + k) = a_{11} + a_{ij} - a_{i1} - a_{1j}.$$

The same happens if we operate on the first column, the ith row or the jth column, while operating on any other row or column clearly does not change E_{ij}. From the hypothesis, we deduce that E_{ij} is divisible by infinitely positive integers N, hence $E_{ij} = 0$. We deduce that

$$a_{11} - a_{1j} = a_{i1} - a_{ij} = c,$$

for all i, $1 < i \leq m$. Adding c to all entries in column j will make this column identical to the first one.

In the same way we can make all columns identical to the first one. Now, it is not difficult to see that operating on rows we can obtain the zero matrix.

6.2 Functions Defined on Sets of Points

Problem 3.14 Let D be the union of $n \geq 1$ concentric circles in the plane. Suppose that the function $f : D \to D$ satisfies

$$d\big(f(A), f(B)\big) \geq d(A, B)$$

for every $A, B \in D$ ($d(M, N)$ is the distance between the points M and N).

Fig. 6.1

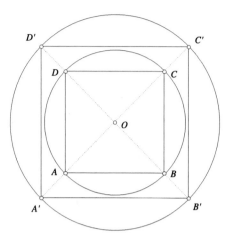

Prove that

$$d\big(f(A), f(B)\big) = d(A, B)$$

for every $A, B \in D$.

Solution Let $D_1, D_2, \ldots, D_n$ be the concentric circles, with radii $r_1 < r_2 < \cdots < r_n$ and center O. We will denote $f(A) = A'$, for an arbitrary point $A \in D$.

We first notice that if $A, B \in D_n$ such that AB is a diameter, then $A'B'$ is also a diameter of D_n. If C is another point on D_n, we have

$$A'C'^2 + B'C'^2 \geq AC^2 + BC^2 = AB^2 = A'B'^2.$$

Because OC' is a median of the triangle $A'B'C'$, it follows that

$$OC'^2 = \frac{1}{2}\big(A'C'^2 + B'C'^2\big) - \frac{1}{4}A'B'^2 = r_n^2,$$

hence $C' \in D_n$ and $A'C' = AC$, $B'C' = BC$. We deduce that $f(D_n) \subset D_n$ and the restriction of f to D_n is an isometry. Now take A, X, Y, Z on D_n such that $AX = AY = A'Z$. It follows that $A'X' = A'Y' = A'Z$, hence one of the points X', Y' coincides with Z. This shows that $f(D_n) = D_n$ and since f is clearly injective it results in the same way that $f(D_i) = D_i$, for all i, $1 \leq i \leq n - 1$, and that all restrictions $f|_{D_i}$ are isometries.

Next we prove that distances between adjacent circles, say D_1 and D_2 are preserved. Take A, B, C, D on D_1 such that $ABCD$ is a square and let A', B', C', D' be the points on D_2 closest to A, B, C, D, respectively.

Then $A'B'C'D'$ is also a square and the distance from A to C' is the maximum between any point on D_1 and any point on D_2. Hence the eight points maintain their relative position under f and this shows that f is an isometry (Fig. 6.1).

Problem 3.15 Let S be a set of $n \geq 4$ points in the plane, such that no three of them are collinear and not all of them lie on a circle. Find all functions $f : S \to \mathbf{R}$ with the property that for any circle C containing at least three points of S,

$$\sum_{P \in C \cap S} f(P) = 0.$$

Solution For two distinct points A, B of S we denote $C_{A,B}$ the set of circles determined by A, B and other points of S. Suppose $C_{A,B}$ has k elements. Since the points of S are not on the same circle, it follows that $k \geq 2$. Because

$$\sum_{P \in C \cap S} f(P) = 0$$

for all $C \in C_{A,B}$, we deduce that

$$\sum_{C \in C_{A,B}} \sum_{P \in C \cap S} f(P) = 0.$$

On the other hand, it is not difficult to see that

$$\sum_{C \in C_{A,B}} \sum_{P \in C \cap S} f(P) = \sum_{P \in S} f(P) + (k-1)\big(f(A) + f(B)\big).$$

Thus the sum $\sum_{P \in S} f(P)$ and $f(A) + f(B)$ have opposite signs, for all A, B in S. If, for instance, $\sum_{P \in S} f(P) \geq 0$, then $f(A) + f(B) \leq 0$, for all A, B in S. Let $S = \{A_1, A_2, \ldots, A_n\}$. Then

$$f(A_1) + f(A_2) \leq 0, \qquad f(A_2) + f(A_3) \leq 0, \ldots, \qquad f(A_n) + f(A_1) \leq 0,$$

yielding

$$2 \sum_{P \in M} f(P) \leq 0$$

hence $\sum_{P \in M} f(P) = 0$ and $f(A) + f(B) = 0$ for all distinct points A, B. This implies that f is the zero function. Indeed, let A, B, C be three distinct points in S. It is not difficult to see that the equalities

$$f(A) + f(B) = 0,$$
$$f(B) + f(C) = 0,$$
$$f(A) + f(C) = 0$$

yield

$$f(A) = f(B) = f(C) = 0$$

and our claim is proved.

Fig. 6.2

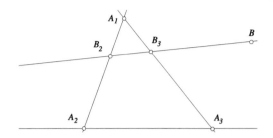

Problem 3.16 Let P be the set of all points in the plane and L be the set of all lines of the plane. Find, with proof, whether there exists a bijective function $f : P \rightarrow L$ such that for any three collinear points A, B, C, the lines $f(A), f(B)$ and $f(C)$ are either parallel or concurrent.

Solution Let $A_i, i = 1, 2, 3$ be three distinct points in the plane and $l_i = f(A_i)$. We claim that if l_1, l_2, l_3 are concurrent or parallel, then A_1, A_2, A_3 are collinear. Indeed, suppose that l_1, l_2, l_3 intersect at M and that $A_1A_2A_3$ is a non-degenerated triangle. Then for any point B in the plane we can find points B_2, B_3 on the lines A_1A_2, A_1A_3, respectively, such that B, B_2, B_3 are collinear (Fig. 6.2).

Because A_1, B_2, A_2 are collinear it follows that $f(B_2)$ is a line passing through M. The same is true for $f(B_1)$, hence also for $f(B)$. This contradicts the surjectivity of f. A similar argument can be given if l_1, l_2, l_3 are parallel.

We conclude that the restriction of f to any line l defines a bijection from l to a pencil of lines (passing through a point or parallel). Consider two pencils P_1 and P_2 of parallel lines. The inverse images of P_1, P_2 are two parallel lines l_1, l_2 (P_1 and P_2 have no common lines, hence l_1 and l_2 have no common points). Let P_3 be a pencil of concurrent lines whose inverse image is a line l, clearly not parallel to l_1, l_2. Let l' be a line parallel to l. Then $f(l')$ is a pencil of concurrent lines and it follows that there is a line through the points corresponding to l and l' whose inverse image would be a point on both l and l', a contradiction. Hence no such functions exists.

Problem 3.17 Let S be the set of interior points of a sphere and C be the set of interior points of a circle. Find, with proof, whether there exists a function $f : S \rightarrow C$ such that $d(A, B) \leq d(f(A), f(B))$, for any points $A, B \in S$.

Solution No such function exists. Indeed, suppose $f : S \rightarrow C$ has the enounced property. Consider a cube inscribed in the sphere and assume with no loss of generality that its sides have length 1. Partition the cube into n^3 smaller cubes and let $A_1, A_2, \ldots, A_{(n+1)^3}$ be their vertices. For all $i \neq j$ we have

$$d(A_i, A_j) \geq \frac{1}{n},$$

hence

$$d\big(f(A_i), f(A_j)\big) \geq \frac{1}{n}.$$

It follows that the disks D_i with centers $f(A_i)$ and radius $\frac{1}{2n}$ are disjoint and contained in a circle C' with radius $r + \frac{1}{n}$, where r is the radius of C. The sum of the areas of these disks is then less than the area of C', hence

$$(n+1)^3 \frac{\pi}{4n^2} \le \pi \left(r + \frac{1}{n} \right)^2.$$

This inequality cannot hold for sufficiently large n, which proves our claim.

Problem 3.18 Let S be the set of all polygons in the plane. Prove that there exists a function $f : S \to (0, +\infty)$ such that

1. $f(P) < 1$, for any $P \in S$;
2. If P_1, $P_2 \in S$ have disjoint interiors and $P_1 \cup P_2 \in S$, then $f(P_1 \cup P_2) = f(P_1) + f(P_2)$.

Solution Consider a covering of the plane with unit squares and denote them by $U_1, U_2, \ldots, U_n, \ldots$. If P is a polygon, define

$$f(P) = \sum_{k \ge 1} \frac{1}{2^k} [P \cap U_k].$$

Because P intersects a finite number of unit squares, the above sum is finite, hence f is well defined. Moreover, f verifies the conditions of the problem. Indeed, we have $[P \cap U_k] \le 1$, for all k and if

$$N = \max\{i \,|\, P \cap U_i \ne \emptyset\}$$

then

$$f(P) = \sum_{k \ge 1}^{N} \frac{1}{2^k} [P \cap U_k] \le \sum_{k \ge 1}^{N} \frac{1}{2^k} = 1 - \frac{1}{2^N} < 1.$$

For the second condition, observe that

$$\left[(P_1 \cap P_2) \cap U_k \right] = [P_1 \cap U_k] + [P_2 \cap U_k]$$

for all polygons P_1, P_2 for which $P_1 \cup P_2 \in S$ and all U_k, hence

$$f(P_1 \cup P_2) = f(P_1) + f(P_2).$$

6.3   Count Twice!

Problem 3.22 Find how many committees with a chairman can be chosen from a set of n persons. Derive the identity

$$\binom{n}{1} + 2\binom{n}{2} + 3\binom{n}{3} + \cdots + n\binom{n}{n} = n2^{n-1}.$$

Solution The number of persons in the committee may vary between 1 and n. Let us count how many such committees number k persons. The k persons can be chosen in $\binom{n}{k}$ ways while the chairman can be chosen in k ways, yielding a total of $k\binom{n}{k}$ committees with k persons. Adding up for $k = 1, 2, \ldots, n$, we see that the total number of committees is

$$\binom{n}{1} + 2\binom{n}{2} + 3\binom{n}{3} + \cdots + n\binom{n}{n}.$$

On the other hand, we can first choose the chairman. This can be done in n ways. Next, we choose the rest of the committee, which is an arbitrary subset of the remaining $n - 1$ persons. Because a set with $n - 1$ elements contains 2^{n-1} subsets, it follows that the committee can be completed in 2^{n-1} ways and the total number of committees is

$$n2^{n-1}.$$

Observation There are many proofs of the given equality. An interesting one is the following. Consider the identity

$$(1 + x)^n = 1 + \binom{n}{1}x + \binom{n}{2}x^2 + \cdots + \binom{n}{n-1}x^{n-1} + \binom{n}{n}x^n.$$

Differentiating both sides with respect to x yields

$$n(1 + x)^{n-1} = \binom{n}{1} + 2\binom{n}{2}x + \cdots + (n-1)\binom{n}{n-1}x^{n-2} + n\binom{n}{n}x^{n-1}.$$

Setting $x = 1$ gives the desired result.

Problem 3.23 In how many ways can one choose k balls from a set containing $n - 1$ red balls and a blue one? Derive the identity

$$\binom{n}{k} = \binom{n-1}{k} + \binom{n-1}{k-1}.$$

Solution We have to choose k balls from a set containing n balls, hence the answer is $\binom{n}{k}$. On the other hand, the blue ball may or may not be among the selected k balls. If the blue ball is selected, then, in fact we have chosen $k - 1$ red balls from $n - 1$ red balls and this can be done in $\binom{n-1}{k-1}$ ways. If the blue ball is not selected, then we have chosen k red balls from $n - 1$ ones. This can be done in $\binom{n-1}{k-1}$ which leads to a total of

$$\binom{n-1}{k} + \binom{n-1}{k-1}$$

possibilities to choose the k balls.

Observation Using similar arguments we can obtain a more general identity. Let us count in how many ways one can choose k balls from a set containing n red and m blue balls. Discarding the color of the balls, the answer is $\binom{n+m}{k}$. If we take into consideration the fact that the chosen k balls can be: all red, or $k-1$ red and 1 blue, or $k-2$ red and 2 blue, etc. we obtain the identity

$$\binom{n+m}{k} = \binom{n}{k}\binom{m}{0} + \binom{n}{k-1}\binom{m}{1} + \binom{n}{k-2}\binom{m}{2} + \cdots + \binom{n}{0}\binom{m}{k}.$$

Problem 3.24 Let S be a set of n persons such that:

 (i) any person is acquainted to exactly k other persons in S;
 (ii) any two persons that are acquainted have exactly l common acquaintances in S;
 (iii) any two persons that are not acquainted have exactly m common acquaintances in S.

Prove that

$$m(n-k) - k(k-l) + k - m = 0.$$

Solution Let a be a fixed element of S. Let us count the triples (a, x, y) such that a, x are acquainted, x, y are acquainted and a, y are not acquainted. Because a is acquainted to exactly k other persons in S, x can be chosen in k ways and for fixed a and x, y can be chosen in $k-1-l$ ways. Thus the number of such triples is

$$k(k-1-l).$$

Let us count again, choosing y first. The number of persons not acquainted to a equals $n-k-1$, hence y can be chosen in $n-k-1$ ways. Because x is a common acquaintance of a and y, it can be chosen in m ways, yielding a total of

$$m(n-k-1)$$

triples. It is not difficult to see that the equality

$$k(k-1-l) = m(n-k-1)$$

is equivalent to the desired one.

Problem 3.25 Let n be an odd integer greater than 1 and let $c_1, c_2, \ldots, c_n$ be integers. For each permutation $a = (a_1, a_2, \ldots, a_n)$ of $\{1, 2, \ldots, n\}$, define

$$S(a) = \sum_{i=1}^{n} c_i a_i.$$

Prove that there exist permutations $a \neq b$ of $\{1, 2, \ldots, n\}$ such that $n!$ is a divisor of $S(a) - S(b)$.

Solution Denote by $\sum_a S(a)$ the sum of $S(a)$ over all $n!$ permutations of $\{1, 2, \ldots, n\}$. We compute $\sum_a S(a) \pmod{n!}$ in two ways. First, assuming that the conclusion is false, it follows that each $S(a)$ has a different remainder mod $n!$, hence these remainders are the numbers $0, 1, 2, \ldots, n! - 1$. It follows that

$$\sum_a S(a) \equiv \frac{1}{2} n!(n! - 1) \pmod{n!}.$$

On the other hand,

$$\sum_a S(a) = \sum_a \sum_{i=1}^n c_i a_i = \sum_{i=1}^n c_i \sum_a a_i.$$

For each i, in $\sum_a a_i$, each of the numbers $1, 2, \ldots, n$, appears $(n - 1)!$ times, hence

$$\sum_a a_i = (n - 1)!(1 + 2 + \cdots + n) = \frac{1}{2}(n + 1)!.$$

It follows that

$$\sum_a S(a) = \frac{1}{2}(n + 1)! \sum_{i=1}^n c_i.$$

We deduce that

$$\frac{1}{2} n!(n! - 1) \equiv \frac{1}{2}(n + 1)! \sum_{i=1}^n c_i \pmod{n!}.$$

Because $n > 1$ is odd, the right-hand side is congruent to $0 \bmod n!$, while the left-hand side is not, a contradiction.

Problem 3.26 Let $a_1 \le a_2 \le \cdots \le a_n = m$ be positive integers. Denote by b_k the number of those a_i for which $a_i \ge k$. Prove that

$$a_1 + a_2 + \cdots + a_n = b_1 + b_2 + \cdots + b_m.$$

Solution Let us consider a $n \times m$ array of numbers (x_{ij}) defined as follows: in row i, the first a_i entries are equal to 1 and the remaining $m - a_i$ entries are equal to 0. For instance, if $n = 3$ and $a_1 = 2, a_2 = 4, a_3 = 5$, the array is

1	1	0	0	0
1	1	1	1	0
1	1	1	1	1

Now, if we examine column j, we notice that the number of 1's in that column equals the number of those a_i greater than or equal to j, hence b_j. The desired result follows by adding up in two ways the 1's in the array. The total number of 1's is $\sum_{i=1}^n a_i$, if counted by rows, and $\sum_{j=1}^m b_j$, if counted by columns.

Problem 3.27 In how many ways can one fill a $m \times n$ table with ± 1 such that the product of the entries in each row and each column equals -1?

Solution Denote by a_{ij} the entries in the table. For $1 \le i \le m-1$ and $1 \le j \le n-1$, we let $a_{ij} = \pm 1$ in an arbitrary way. This can be done in $2^{(m-1)(n-1)}$ ways. The values for a_{mj} with $1 \le j \le n-1$ and for a_{in}, with $1 \le i \le m-1$ are uniquely determined by the condition that the product of the entries in each row and each column equals -1. The value of a_{mn} is also uniquely determined but it is necessary that

$$\prod_{j=1}^{n-1} a_{mj} = \prod_{i=1}^{m-1} a_{in}. \qquad (*)$$

If we denote

$$P = \prod_{i=1}^{m-1} \prod_{j=1}^{n-1} a_{ij}$$

we observe that

$$P \prod_{j=1}^{n-1} a_{mj} = (-1)^{n-1}$$

and

$$P \prod_{i=1}^{m-1} a_{in} = (-1)^{m-1}$$

hence $(*)$ holds if and only if m and n have the same parity.

Problem 3.28 Let n be a positive integer. Prove that

$$\sum_{k=0}^{n} \binom{n}{k} \binom{n+k}{k} = \sum_{k=0}^{n} 2^k \binom{n}{k}^2.$$

Solution Let us count in two ways the number of ordered pairs (A, B), where A is a subset of $\{1, 2, \ldots, n\}$, and B is a subset of $\{1, 2, \ldots, 2n\}$ with n elements and disjoint from A.

First, for $0 \le k \le n$, choose a subset A of $\{1, 2, \ldots, n\}$ having $n-k$ elements. This can be done in

$$\binom{n}{n-k} = \binom{n}{k}$$

ways. Next, choose B, a subset with n elements of $\{1, 2, \ldots, 2n\} - A$. Since $\{1, 2, \ldots, 2n\} - A$ has $2n - (n-k) = n+k$ elements, B can be chosen in

$$\binom{n+k}{n} = \binom{n+k}{k}$$

ways. We deduce, by adding on k, that the number of such pairs equals

$$\sum_{k=0}^{n} \binom{n}{k}\binom{n+k}{k}.$$

On the other hand, we could start by choosing the subsets $B' \subset \{1, 2, \ldots, n\}$ and $B'' \subset \{n+1, n+2, \ldots, 2n\}$, both with k elements, and define

$$B = B'' \cup (\{1, 2, \ldots, n\} - B').$$

Since for given k each of the sets B' and B'' can be chosen in $\binom{n}{k}$ ways, B can be chosen in $\binom{n}{k}^2$ ways.

Finally, pick A to be an arbitrary subset of B'. There are 2^k ways to do this. Adding on k, we obtain that the total number of pairs (A, B) equals

$$\sum_{k=0}^{n} 2^k \binom{n}{k}^2,$$

hence the conclusion.

Problem 3.29 Prove that

$$1^2 + 2^2 + \cdots + n^2 = \binom{n+1}{2} + 2\binom{n+1}{3}.$$

Solution Let us count the number of ordered triples of integers (a, b, c) satisfying $0 \le a, b < c \le n$.

For fixed c, a and b can independently take values in the set $\{0, 1, \ldots, c-1\}$, hence there are c^2 such triples. Since c can take any integer value between 1 and n, the total number of triples equals

$$1^2 + 2^2 + \cdots + n^2.$$

On the other hand, there are $\binom{n+1}{2}$ triples of the form (a, a, c) and $2\binom{n+1}{3}$ triples (a, b, c) with $a \ne b$. The latter results from the following argument: we can chose three distinct elements from $\{0, 1, \ldots, n\}$ in $\binom{n+1}{3}$ ways. With these three elements, say $x < y < z$, we can form two triples satisfying the required condition: (x, y, z) and (y, x, z).

Problem 3.30 Let n and k be positive integers and let S be a set of n points in the plane such that

(a) no three points of S are collinear, and
(b) for every point P of S there are at least k points of S equidistant from P.

Prove that

$$k < \frac{1}{2} + \sqrt{2n}.$$

Solution Let $P_1, P_2, \ldots, P_n$ be the given points. Let us estimate the number of isosceles triangles whose vertices lie in S.

For each P_i, there are at least $\binom{k}{2}$ triangles $P_i P_j P_k$, with $P_i P_j = P_i P_k$, hence the total number of isosceles triangles is at least $n\binom{k}{2}$.

For each pair (P_i, P_j), with $i \neq j$, there exist at most two points in S equidistant from P_i and P_j. This is because all such points lie on the perpendicular bisector of the line segment $P_i P_j$ and no three points of S are collinear. Thus, the total number of isosceles triangles is at most $2\binom{n}{2}$. We deduce that

$$n\binom{k}{2} \leq 2\binom{n}{2},$$

which simplifies to

$$2(n-1) \geq k(k-1).$$

Suppose, by way of contradiction, that

$$k \geq \frac{1}{2} + \sqrt{2n}.$$

Then

$$k(k-1) \geq \left(\sqrt{2n} + \frac{1}{2}\right)\left(\sqrt{2n} - \frac{1}{2}\right) = 2n - \frac{1}{4} > 2(n-1),$$

a contradiction.

Problem 3.31 Prove that

$$\tau(1) + \tau(2) + \cdots + \tau(n) = \left\lfloor \frac{n}{1} \right\rfloor + \left\lfloor \frac{n}{2} \right\rfloor + \cdots + \left\lfloor \frac{n}{n} \right\rfloor,$$

where $\tau(k)$ denotes the number of divisors of the positive integer k.

Solution Observe that the integer a, with $1 \leq a \leq n$, is counted by $\tau(k)$ if and only if a is a divisor of k, or, equivalently, if k is a multiple of a. Thus, the number of times a is counted in the left-hand side of our equality equals the number of multiples of a in the set $\{1, 2, \ldots, n\}$. But this number obviously equals $\lfloor \frac{n}{a} \rfloor$. Adding on a gives the desired result.

Problem 3.32 Prove that

$$\sigma(1) + \sigma(2) + \cdots + \sigma(n) = \left\lfloor \frac{n}{1} \right\rfloor + 2\left\lfloor \frac{n}{2} \right\rfloor + \cdots + n\left\lfloor \frac{n}{n} \right\rfloor,$$

where $\sigma(k)$ denotes the sum of divisors of the positive integer k.

Solution The solution is similar to the previous one. The integer a, with $1 \le a \le n$, is a term of the sum $\sigma(k)$ if and only if a is a divisor of k. There are $\lfloor \frac{n}{a} \rfloor$ multiples of a in the set $\{1, 2, \ldots, n\}$ and therefore the sum of all divisors equal to a is $a\lfloor \frac{n}{a} \rfloor$.

Problem 3.33 Prove that

$$\varphi(1)\left\lfloor \frac{n}{1} \right\rfloor + \varphi(2)\left\lfloor \frac{n}{2} \right\rfloor + \cdots + \varphi(n)\left\lfloor \frac{n}{n} \right\rfloor = \frac{n(n+1)}{2},$$

where φ denotes Euler's totient function.

Solution Observe that the right-hand side can be written as

$$\frac{n(n+1)}{2} = 1 + 2 + \cdots + n.$$

Using the result of Problem 3.21 of Sect. 3.3, we obtain

$$\frac{n(n+1)}{2} = \sum_{d|1} \varphi(d) + \sum_{d|2} \varphi(d) + \cdots + \sum_{d|n} \varphi(d).$$

Now, for some k, $1 \le k \le n$, let us count how many times $\varphi(k)$ appears in right-hand side of the above equality. Clearly, $\varphi(k)$ is a term of the sum $\sum_{d|m} \varphi(d)$ if and only if k is a divisor of m, or, equivalently, m is a multiple of k. So, $\varphi(k)$ appears as many times as the number of multiples of k in the set $\{1, 2, \ldots, n\}$, that is, $\lfloor \frac{n}{k} \rfloor$.

We conclude that the sum can be written as

$$\varphi(1)\left\lfloor \frac{n}{1} \right\rfloor + \varphi(2)\left\lfloor \frac{n}{2} \right\rfloor + \cdots + \varphi(n)\left\lfloor \frac{n}{n} \right\rfloor,$$

giving the required result.

Alternatively, we could count in two ways the number of fractions $\frac{a}{b}$ with $1 \le a \le b \le n$ where we do not insist that our fraction be in lowest terms, so for example $\frac{1}{2}$ and $\frac{2}{4}$ would count as different fractions. First, this number is clearly

$$\sum_{b=1}^{n} b = \frac{n(n+1)}{2}$$

since this is the number of such pairs (a, b). Second, we count fractions based on their reduced forms. We saw in Problem 3.21 that the number of reduced fractions with denominator d is $\phi(d)$. The number of multiples of such a fraction with denominator at most n is $\lfloor n/d \rfloor$ so the number is also

$$\sum_{d=1}^{n} \phi(d)\left\lfloor \frac{n}{d} \right\rfloor.$$

6.4 Sequences of Integers

Problem 3.37 Prove that there exist sequences of odd integers $(x_n)_{n \geq 3}, (y_n)_{n \geq 3}$ such that

$$7x_n^2 + y_n^2 = 2^n$$

for all $n \geq 3$.

Solution For $n = 3$, define $x_3 = y_3 = 1$. Suppose that for $n \geq 3$ there exist odd integers x_n and y_n such that $7x_n^2 + y_n^2 = 2^n$. Observe that the integers

$$\frac{x_n + y_n}{2} \quad \text{and} \quad \frac{x_n - y_n}{2}$$

cannot be both even, since their sum is odd.

If $\frac{x_n + y_n}{2}$ is odd, we define

$$x_{n+1} = \frac{x_n + y_n}{2}, \qquad y_{n+1} = \frac{7x_n - y_n}{2}$$

and the conclusion follows by noticing that

$$7x_{n+1}^2 + y_{n+1}^2 = \frac{1}{4}\left(7(x_n + y_n)^2 + (7x_n - y_n)^2\right) = 2\left(7x_n^2 + y_n^2\right) = 2^{n+1}.$$

If $\frac{x_n - y_n}{2}$ is odd, we define

$$x_{n+1} = \frac{x_n - y_n}{2}, \qquad y_{n+1} = \frac{7x_n + y_n}{2}$$

and a similar computation yields the result.

Problem 3.38 Let $x_1 = x_2 = 1, x_3 = 4$ and

$$x_{n+3} = 2x_{n+2} + 2x_{n+1} - x_n$$

for all $n \geq 1$. Prove that x_n is a square for all $n \geq 1$.

Solution We first notice that $x_1 = x_2 = 1^2, x_3 = 2^2, x_4 = 3^2, x_5 = 5^2, x_6 = 8^2$, and so forth. This leads to the assumption that $x_n = F_n^2$, where F_n is the nth term of the Fibonacci sequence defined by $F_1 = F_2 = 1$ and $F_{n+1} = F_n + F_{n-1}$, for all $n \geq 2$. We prove this assertion inductively. Suppose it is true for all $k \leq n + 2$. Then

$$x_{n+3} = 2F_{n+2}^2 + 2F_{n+1}^2 - F_n^2 = 2F_{n+2}^2 + 2F_{n+1}^2 - (F_{n+2} - F_{n+1})^2$$

$$= F_{n+2}^2 + F_{n+1}^2 + 2F_{n+1}F_{n+2} = (F_{n+2} + F_{n+1})^2 = F_{n+3}^2,$$

and the assertion is proved.

Problem 3.39 The sequence $(a_n)_{n\geq 0}$ is defined by $a_0 = a_1 = 1$ and

$$a_{n+1} = 14a_n - a_{n-1}$$

for all $n \geq 1$. Prove that the number $2a_n - 1$ is a square for all $n \geq 0$.

Solution We have $2a_0 - 1 = 1, 2a_1 - 1 = 1, 2a_2 - 1 = 5^2, 2a_3 - 1 = 19^2, 2a_4 - 1 = 71^2$. We observe that if we define $b_0 = -1, b_1 = 1, b_2 = 5, b_3 = 19, b_4 = 71$, then $b_{n+1} = 4b_n - b_{n-1}$ for $1 \leq n \leq 3$. We will prove inductively that

$$2a_n - 1 = b_n^2,$$

where $b_0 = -1, b_1 = 1$ and $b_{n+1} = 4b_n - b_{n-1}$ for all $n \geq 1$. Suppose this is true for $1, 2, \ldots, n$ and observe that

$$\begin{aligned}
2a_{n+1} - 1 &= 14(2a_n - 1) - (2a_{n-1} - 1) + 12 = 14b_n^2 - b_{n-1}^2 + 12 \\
&= 16b_n^2 - 8b_n b_{n-1} + b_{n-1}^2 - 2b_n^2 + 8b_n b_{n-1} - 2b_{n-1}^2 + 12 \\
&= (4b_n - b_{n-1})^2 - 2(b_n^2 + b_{n-1}^2 - 4b_n b_{n-1} - 6) \\
&= b_{n+1}^2 - 2(b_n^2 + b_{n-1}^2 - 4b_n b_{n-1} - 6).
\end{aligned}$$

Hence it suffices to prove that $b_n^2 + b_{n-1}^2 - 4b_n b_{n-1} - 6 = 0$. This follows also by induction. It is true for $n = 1$ and $n = 2$. Suppose it holds for n and observe that

$$\begin{aligned}
b_{n+1}^2 + b_n^2 - 4b_{n+1}b_n - 6 &= (4b_n - b_{n-1})^2 + b_n^2 - 4(4b_n - b_{n-1})b_n - 6 \\
&= b_n^2 + b_{n-1}^2 - 4b_n b_{n-1} - 6 = 0,
\end{aligned}$$

as desired.

Problem 3.40 The sequence $(x_n)_{n\geq 1}$ is defined by $x_1 = 0$ and

$$x_{n+1} = 5x_n + \sqrt{24x_n^2 + 1}$$

for all $n \geq 1$. Prove that all x_n are positive integers.

Solution We first notice that the sequence is increasing and all of its terms are positive. Next we observe that the recursive relation is equivalent to

$$x_{n+1}^2 - 10x_n x_{n+1} + x_n^2 - 1 = 0.$$

Replacing n by $n - 1$ yields

$$x_n^2 - 10x_n x_{n-1} + x_{n-1}^2 - 1 = 0,$$

hence for $n \geq 2$ the numbers x_{n+1} and x_{n-1} are distinct roots of the equation

$$x^2 - 10xx_n + x_n^2 - 1 = 0.$$

The Viète's relations yield $x_{n+1} + x_{n-1} = 10x_n$, or

$$x_{n+1} = 10x_n - x_{n-1}$$

for all $n \geq 2$. Because $x_1 = 1$ and $x_2 = 10$, it follows inductively that all x_n are positive integers.

Problem 3.41 Let $(a_n)_{n \geq 1}$ be an increasing sequence of positive integers such that

1. $a_{2n} = a_n + n$ for all $n \geq 1$;
2. if a_n is a prime, then n is a prime.

Prove that $a_n = n$, for all $n \geq 1$.

Solution Let $a_1 = c$. Then $a_2 = a_1 + 1 = c + 1, a_4 = a_2 + 2 = c + 3$. Since the sequence is increasing, it follows that $a_3 = c + 2$. We prove that $a_n = c + n - 1$ for all $n \geq 1$. Indeed, if $n = 2^k$ for some integer k this follows by induction on k. Suppose that

$$a_{2^k} = c + 2^k - 1.$$

Then

$$a_{2^{k+1}} = a_{2 \cdot 2^k} = a_{2^k} + 2^k = c + 2^{k+1} - 1.$$

If $2^k < n < 2^{k+1}$, then

$$c + 2^k - 1 = a_{2^k} < a_{2^k + 1} < \cdots < a_n < \cdots < a_{2^{k+1}} = c + 2^{k+1} - 1$$

and this is possible only if $a_n = c + n - 1$.

Next we prove that $c = 1$. Suppose that $c \geq 2$ and let $p < q$ be two consecutive prime numbers greater than c. We have

$$a_{q-c+1} = c + q - c = q,$$

hence $q - c + 1$ is a prime and clearly $q - c + 1 \leq p$. It follows that for any consecutive prime numbers $p < q$ we have

$$q - p \leq c - 1.$$

The numbers $(c + 1)! + 2, (c + 1)! + 3, \ldots, (c + 1)! + c + 1$ are all composite, hence if p and q are the consecutive primes such that

$$p < (c + 1)! + 2 < (c + 1)! + c + 1 < q$$

then

$$q - p > c - 1,$$

a contradiction. It follows that $c = 0$ and $a_n = n$ for all $n \geq 1$.

Problem 3.42 Let $a_0 = a_1 = 1$ and $a_{n+1} = 2a_n - a_{n-1} + 2$, for all $n \geq 1$. Prove that

$$a_{n^2+1} = a_{n+1}a_n,$$

for all $n \geq 0$.

Solution We will find a closed form for a_n. Denoting $b_n = a_{n+1} - a_n$, we observe that the given recursive equation becomes

$$b_n = b_{n-1} + 2,$$

thus $(b_n)_{n \geq 0}$ is an arithmetic sequence and hence $b_n = b_0 + 2n = 2n$. Writing $a_{k+1} - a_k = 2(k-1)$ for $k = 1, 2, \ldots, n-1$ and adding up yield $a_n = n^2 - n + 1$, for all $n \geq 0$. Now, an elementary computation shows that

$$a_{n^2+1} = \left(n^2 + 1\right)^2 - \left(n^2 + 1\right) + 1 = \left(n^2 + n + 1\right)\left(n^2 - n + 1\right) = a_{n+1}a_n,$$

as desired.

Problem 3.43 Let $a_0 = 1$ and $a_{n+1} = a_0 \cdots a_n + 4$, for all $n \geq 0$. Prove that

$$a_n - \sqrt{a_{n+1}} = 2,$$

for all $n \geq 1$.

Solution We will prove the equivalent statement

$$a_{n+1} = (a_n - 2)^2.$$

We have

$$
\begin{aligned}
a_{n+1} &= a_0 \cdots a_{n-1} \cdot a_n + 4 \\
&= (a_n - 4) \cdot a_n + 4 \\
&= a_n^2 - 4a_n + 4 \\
&= (a_n - 2)^2,
\end{aligned}
$$

which concludes the proof.

Problem 3.44 The sequence $(x_n)_{n \geq 1}$ is defined by $x_1 = 1, x_2 = 3$ and $x_{n+2} = 6x_{n+1} - x_n$, for all $n \geq 1$. Prove that $x_n + (-1)^n$ is a perfect square, for all $n \geq 1$.

Solution Let $y_n = x_n + (-1)^n$. By inspection, we find that $y_1 = 0, y_2 = 4, y_3 = 16, y_4 = 100, y_5 = 576$ etc. Denote by $z_n = \sqrt{y_n}$; then $z_1 = 0, z_2 = 2, z_3 = 4, z_4 = 10, z_5 = 24$, and so on. The first terms of the sequence $(z_n)_{n \geq 1}$ suggest that $z_{n+2} = 2z_{n+1} + z_n$, for all $n \geq 1$. Indeed, we will prove by induction the following statement: $y_n = z_n^2$, where $z_1 = 0, z_2 = 2$ and $z_{n+2} = 2z_{n+1} + z_n$, for all $n \geq 1$.

The base case is easy to deal with. In order to prove the inductive step, we need a recursive equation for the sequence $(y_n)_{n \geq 1}$. Since $x_n = y_n - (-1)^n$, we have

$$y_{n+2} - (-1)^{n+2} = 6y_{n+1} - 6(-1)^{n+1} - y_n + (-1)^n,$$

and also

$$y_{n+1} - (-1)^{n+1} = 6y_n - 6(-1)^n - y_{n-1} + (-1)^{n-1}.$$

Adding up yields

$$y_{n+2} + y_{n+1} = 6y_{n+1} + 6y_n - y_n - y_{n-1},$$

or

$$y_{n+2} = 5y_{n+1} + 5y_n - y_{n-1},$$

for all $n \geq 2$. Now, we are ready to prove the inductive step. Assume $y_k = z_k^2$, for $k = 1, 2, \ldots, n + 1$. Then

$$
\begin{aligned}
y_{n+2} &= 5z_{n+1}^2 + 5z_n^2 - z_{n-1}^2 \\
&= 5z_{n+1}^2 + 5z_n^2 - (z_{n+1} - 2z_n)^2 \\
&= 4z_{n+1}^2 + z_n^2 + 4z_{n+1}z_n \\
&= (2z_{n+1} + z_n)^2 \\
&= z_{n+2}^2,
\end{aligned}
$$

and we are done.

Observation The informed reader may notice that an alternative solution is possible if we write x_n in closed form. Indeed, it is known that if α and β, with $\alpha \neq \beta$, are the roots of the quadratic equation

$$x^2 = ax + b,$$

then any sequence satisfying

$$x_{n+2} = ax_{n+1} + bx_n,$$

for all $n \geq 1$, has the form

$$x_n = c_1 \alpha^n + c_2 \beta^n,$$

where the constants c_1 and c_2 can be determined from the first two terms of the sequence. In our case, the roots of the equation

$$x^2 = 6x - 1$$

are

$$\alpha = 3 + 2\sqrt{2}, \qquad \beta = 3 - 2\sqrt{2},$$

hence

$$x_n = c_1(3 + 2\sqrt{2})^n + c_2(3 - 2\sqrt{2})^n.$$

Since $x_1 = 1$ and $x_2 = 3$, we obtain

$$c_1 = \frac{1}{2}(3 - 2\sqrt{2}), \qquad c_2 = \frac{1}{2}(3 + 2\sqrt{2}),$$

hence

$$x_n = \frac{1}{2}\left((3 + 2\sqrt{2})^{n-1} + (3 - 2\sqrt{2})^{n-1}\right),$$

for all $n \geq 1$. It follows that

$$x_n + (-1)^n = \frac{1}{2}\left((3 + 2\sqrt{2})^{n-1} + (3 - 2\sqrt{2})^{n-1} + 2(-1)^n\right)$$

$$= \frac{1}{2}\left((1 + \sqrt{2})^{2(n-1)} + (1 - \sqrt{2})^{2(n-1)} - 2(-1)^{n-1}\right)$$

$$= \left(\frac{(1 + \sqrt{2})^{n-1} - (1 - \sqrt{2})^{n-1}}{\sqrt{2}}\right)^2.$$

The sequence

$$w_n = \frac{(1 + \sqrt{2})^{n-1} - (1 - \sqrt{2})^{n-1}}{\sqrt{2}}$$

has the form

$$w_n = c_1\alpha^n + c_2\beta^n,$$

with

$$\alpha = 1 + \sqrt{2}, \qquad \beta = 1 - \sqrt{2},$$

hence it satisfies the recursive equation

$$w_{n+2} = 2w_{n+1} + w_n,$$

for all $n \geq 1$. Finally, since $w_1 = 0$ and $w_1 = 2$, we deduce that w_n is an integer for all $n \geq 1$.

Problem 3.45 Let $(a_n)_{n\geq 1}$ be a sequence of non-negative integers such that $a_n \geq a_{2n} + a_{2n+1}$, for all $n \geq 1$. Prove that for any positive integer N we can find N consecutive terms of the sequence, all equal to zero.

Solution Let us first show that at least one term of the sequence equals zero. Suppose the contrary, that is, all terms are positive integers. But then

$$a_1 \geq a_2 + a_3 \geq a_4 + a_5 + a_6 + a_7 \geq \cdots \geq a_{2^n} + a_{2^n+1} + \cdots + a_{2^{n+1}-1} \geq 2^n,$$

for all $n \geq 1$, which is absurd. Thus, at least one term, say a_k, equals zero. But then

$$0 = a_k \geq a_{2k} + a_{2k+1} \geq \cdots \geq a_{2^n k} + a_{2^n k+1} + \cdots + a_{2^n k+2^n-1},$$

hence $a_{2^n k} = a_{2^n k+1} = \cdots = a_{2^n k+2^n-1} = 0$. We found 2^n consecutive terms of our sequence, all equal to zero, which clearly proves the claim.

Observation An nontrivial example of such a sequence is the following:

$$a_n = \begin{cases} 1, & \text{if } n = 2^k, \text{ for some integer } k, \\ 0, & \text{otherwise.} \end{cases}$$

6.5 Equations with Infinitely Many Solutions

Problem 3.48 Find all triples of integers (x, y, z) such that

$$x^2 + xy = y^2 + xz.$$

Solution The given equation is equivalent to

$$x(x - z) = y(y - x).$$

Denote by $d = \gcd(x, y)$. Then $x = da$, $y = db$, with $\gcd(a, b) = 1$. We deduce that $y - x = ka$ and $x - z = kb$ for some integer k. Because $\gcd(a, b - a) = \gcd(a, b) = 1$, it follows that $b - a$ divides k. Setting $k = m(b - a)$ we obtain $d = ma$ and the solutions are

$$x = ma^2, \qquad y = mab, \qquad z = m(a^2 + ab - b^2)$$

where m, a, b are arbitrary integers.

Problem 3.49 Let n be an integer number. Prove that the equation

$$x^2 + y^2 = n + z^2$$

has infinitely many integer solutions.

Solution The equation is equivalent to

$$(x - z)(x + z) + y^2 = n.$$

If we set $x - z = 1$, then we obtain

$$2x - 1 + y^2 = n$$

and

$$x = \frac{n + 1 - y^2}{2}.$$

Now, it suffices to take $y = n + m$, where m is an odd integer to insure that x is an integer as well. Indeed, if $m = 2k + 1$, then

$$\frac{n + 1 - y^2}{2} = \frac{n + 1 - (n + 2k + 1)^2}{2} = -\frac{n(n + 1)}{2} - 2nk - 2k^2 - 2k,$$

obviously an integer. Since $z = x - 1$, it is also an integer number.

Problem 3.50 Let m be a positive integer. Find all pairs of integers (x, y) such that

$$x^2(x^2 + y) = y^{m+1}.$$

Solution Multiplying the equation by 4 and adding y^2 to both sides yields the equivalent form

$$(2x^2 + y)^2 = y^2 + 4y^{m+1}$$

or

$$(2x^2 + y)^2 = y^2(1 + 4y^{m-1}).$$

It follows that $1 + 4y^{m-1}$ is an odd square, say $(2a + 1)^2$. We obtain $y^{m-1} = a(a + 1)$ and since a and $a + 1$ are relatively prime integers, each of them must be the $(m - 1)$th power of some integers. Clearly, this is possible only if $m = 2$, hence $y = a(a + 1)$. It follows that

$$2x^2 + a(a + 1) = a(a + 1)(2a + 1)$$

hence $x^2 = a^2(a + 1)$. We deduce that $a + 1$ is a square and setting $a + 1 = t^2$ yields $x = t^3 - t$ and $y = t^4 - t^2$.

Problem 3.51 Let m be a positive integer. Find all pairs of integers (x, y) such that

$$x^2(x^2 + y^2) = y^{m+1}.$$

Solution The equation can be written in the equivalent form

$$(2x^2 + y)^2 = y^4 + 4y^{m+1}.$$

Observe that $y^4 + 4y^{m+1}$ cannot be a square for $m = 1$. Indeed, in this case

$$y^4 + 4y^{m+1} = y^4 + 4y^2 = y^2(y^2 + 4)$$

and no squares differ by 4. A similar argument works for $m = 2$.

Now, for $m \geq 3$, we write the equation in the equivalent form

$$\left(2x^2 + y\right)^2 = y^4\left(1 + 4y^{m-3}\right).$$

As in the previous solution we deduce that $y = a(a + 1)$ for some integer a, then $x = a^3(a + 1)$. It follows that $a = t^2$ for some integer t and the solutions are $x = t^5 + t^3$, $y = t^4 + t^2$.

Problem 3.52 Find all non-negative integers a, b, c, d, n such that

$$a^2 + b^2 + c^2 + d^2 = 7 \cdot 4^n.$$

Solution For $n = 0$, we have $2^2 + 1^2 + 1^2 + 1^2 = 7$, hence $(a, b, c, d) = (2, 1, 1, 1)$ and all permutations. If $n \geq 1$, then $a^2 + b^2 + c^2 + d^2 \equiv 0 \pmod 4$, hence the numbers have the same parity. We analyze two cases.

(a) The numbers a, b, c, d are odd. We write $a = 2a' + 1$, etc. We obtain

$$4a'(a' + 1) + 4b'(b' + 1) + 4c'(c' + 1) + 4d'(d' + 1) = 4\left(7 \cdot 4^{n-1} - 1\right).$$

The left-hand side of the equality is divisible by 8, hence $7 \cdot 4^{n-1} - 1$ must be even. This happens only for $n = 1$. We obtain $a^2 + b^2 + c^2 + d^2 = 28$, with the solutions $(3, 3, 3, 1)$ and $(1, 1, 1, 5)$.

(b) The numbers a, b, c, d are even. Write $a = 2a'$, etc. We obtain

$$a'^2 + b'^2 + c'^2 + d'^2 = 7 \cdot 4^{n-1},$$

so we proceed recursively.

Finally, we obtain the solutions $(2^{n+1}, 2^n, 2^n, 2^n)$, $(3 \cdot 2^n, 3 \cdot 2^n, 3 \cdot 2^n, 2^n)$, $(2^n, 2^n, 2^n, 5 \cdot 2^n)$, and the respective permutations.

Problem 3.53 Show that there are infinitely many systems of positive integers (x, y, z, t) which have no common divisor greater than 1 and such that

$$x^3 + y^3 + z^2 = t^4.$$

Solution Consider the identity

$$(a + 1)^4 - (a - 1)^4 = 8a^3 + 8a.$$

Taking $a = b^3$, with b an even integer gives

$$\left(b^3 + 1\right)^4 = \left(2b^3\right)^3 + (2b)^3 + \left(\left(b^3 - 1\right)^2\right)^2.$$

Since b is even, $b^3 + 1$ and $b^3 - 1$ are odd integers. It follows that the numbers $x = 2b^3$, $y = 2b$, $z = (b^3 - 1)^2$ and $t = b^3 + 1$ have no common divisor greater than 1.

Problem 3.54 Let $k \geq 6$ be an integer number. Prove that the system of equations

$$\begin{cases} x_1 + x_2 + \cdots + x_{k-1} = x_k, \\ x_1^3 + x_2^3 + \cdots + x_{k-1}^3 = x_k, \end{cases}$$

has infinitely many integral solutions.

Solution Consider the identity

$$(m+1)^3 + (m-1)^3 + (-m)^3 + (-m)^3 = 6m.$$

If n is an arbitrary integer, then $\frac{n-n^3}{6}$ is also an integer since $n - n^3 = -(n-1)n \times (n+1)$ and the product of three consecutive integers is divisible by 6. Setting $m = \frac{n-n^3}{6}$ in the identity above gives

$$\left(\frac{n-n^3}{6} + 1 \right)^3 + \left(\frac{n-n^3}{6} - 1 \right)^3 + \left(\frac{n^3-n}{6} \right)^3 + \left(\frac{n^3-n}{6} \right)^3 + n^3 = n.$$

On the other hand, we have

$$\left(\frac{n-n^3}{6} + 1 \right) + \left(\frac{n-n^3}{6} - 1 \right) + \frac{n^3-n}{6} + \frac{n^3-n}{6} + n = n,$$

yielding the following solution for $k = 6$: $x_1 = \frac{n-n^3}{6} + 1$, $x_2 = \frac{n-n^3}{6} - 1$, $x_3 = x_4 = \frac{n^3-n}{6}$, $x_5 = x_6 = n$. For $k > 6$ we can take x_1 to x_5 as before, $x_6 = -n$ and $x_i = 0$ for all $i > 6$.

Problem 3.55 Solve in integers the equation

$$x^2 + y^2 = (x - y)^3.$$

Solution If we denote $a = x - y$ and $b = x + y$, the equation rewrites as

$$a^2 + b^2 = 2a^3,$$

or

$$2a - 1 = \frac{b^2}{a^2}.$$

We see that $2a - 1$ is the square of a rational number, hence it is the square of an (odd) integer. Let $2a - 1 = (2n+1)^2$. It follows that $a = 2n^2 + 2n + 1$ and $b = a(2n+1) = (2n+1)(2n^2 + 2n + 1)$. Finally, we obtain

$$x = 2n^3 + 4n^2 + 3n + 1, \qquad y = 2n^3 + 2n^2 + n,$$

where n is an arbitrary integer number.

Problem 3.56 Let a and b be positive integers. Prove that if the equation

$$ax^2 - by^2 = 1$$

has a solution in positive integers, then it has infinitely many solutions.

Solution Factor the left-hand side to obtain

$$\left(x\sqrt{a} - y\sqrt{b}\right)\left(x\sqrt{a} + y\sqrt{b}\right) = 1.$$

Cubing both sides yields

$$\left(\left(x^3 a + 3xy^2 b\right)\sqrt{a} - \left(3x^2 ya + y^3 b\right)\sqrt{b}\right)\left(\left(x^3 a + 3xy^2 b\right)\sqrt{a}\right.$$
$$\left. + \left(3x^2 ya + y^3 b\right)\sqrt{b}\right) = 1.$$

Multiplying out, we obtain

$$a\left(x^3 a + 3xy^2 b\right)^2 - b\left(3x^2 ya + y^3 b\right)^2 = 1.$$

Therefore, if (x_1, y_1) is a solution of the equation, so is (x_2, y_2), with $x_2 = x_1^3 a + 3x_1 y_1^2 b$, and $y_2 = 3x_1^2 y_1 a + y_1^3 b$. Clearly, $x_2 > x_1$ and $y_2 > y_1$. Continuing in this way we obtain infinitely many solutions in positive integers.

Problem 3.57 Prove that the equation

$$\frac{x+1}{y} + \frac{y+1}{x} = 4$$

has infinitely many solutions in positive integers.

Solution Suppose that the equation has a solution (x_1, y_1) with $x_1 \leq y_1$. Clearing denominators, we can write the equation under the form

$$x^2 - (4y - 1)x + y^2 + y = 0,$$

that is, a quadratic in x. One of the roots is x_1, therefore, by Vieta's theorem, the second one is $4y_1 - 1 - x_1$. Observe that $4y_1 - 1 - x_1 \geq 4x_1 - 1 - x_1 = 3x_1 - 1 \geq 2x_1 > 0$, hence $4y_1 - 1 - x_1$ is a positive integer.

It follows that $(4y_1 - 1 - x_1, y_1)$ is another solution of the system. Because the equation is symmetric, we obtain that $(x_2, y_2) = (y_1, 4y_1 - 1 - x_1)$ is also a solution. To end the proof, observe that $x_2 + y_2 = 5y_1 - 1 - x_1 > x_1 + y_1$ and that $(1, 1)$ is a solution. Thus, we can generate infinitely many solutions:

$$(1, 1) \to (1, 2) \to (2, 6) \to (6, 21) \to \cdots .$$

Problem 3.58 Prove that the equation

$$x^3 + y^3 - 2z^3 = 6(x + y + 2z)$$

has infinitely many solutions in positive integers.

Solution Observe that

$$(n + 1)^3 + (n - 1)^3 - 2n^3 = 6n,$$

and

$$(n + 1) + (n - 1) + 2n = 4n.$$

Taking $x = 2(n + 1)$, $y = 2(n - 1)$, and $z = 2n$, we see that

$$x^3 + y^3 - 2z^3 = 48n,$$

and

$$6(x + y + 2z) = 48n.$$

Thus, the triples $(x, y, z) = (2(n + 1), 2(n - 1, 2n))$ are solutions in positive integers for all integers $n > 1$.

6.6 Equations with No Solutions

Problem 3.62 Prove that the equation

$$4xy - x - y = z^2$$

has no positive integer solutions.

Solution We write the equation in the equivalent form

$$(4x - 1)(4y - 1) = 4z^2 + 1.$$

Let p be a prime divisor of $4x - 1$. Then

$$4z^2 + 1 \equiv 0 \pmod{p}$$

or

$$(2z)^2 \equiv -1 \pmod{p}.$$

On the other hand, Fermat's theorem yields

$$(2z)^{p-1} \equiv 1 \pmod{p}$$

hence

$$(2z)^{p-1} \equiv \left(2z^2\right)^{\frac{p-1}{2}} \equiv (-1)^{\frac{p-1}{2}} \equiv 1 \ (\mathrm{mod}\ p).$$

This implies that $p \equiv 1 \ (\mathrm{mod}\,4)$. It follows that all prime divisors of $4x - 1$ are congruent to 1 modulo 4, hence $4x - 1 \equiv 1 \ (\mathrm{mod}\,4)$, a contradiction.

Problem 3.63 Prove that the equation

$$6\left(6a^2 + 3b^2 + c^2\right) = 5d^2$$

has no solution in non-zero integers.

Solution We can assume that $\gcd(a, b, c, d) = 1$, otherwise we simplify the equation with a suitable integer. Clearly, d is divisible by 6, so let $d = 6m$. We obtain

$$6a^2 + 3b^2 + c^2 = 30m^2,$$

hence c is divisible by 3. Replacing $c = 3n$ in the equation yields

$$2a^2 + b^2 + 3n^2 = 10m^2.$$

If b and n are odd, then $b^2 \equiv 1 \ (\mathrm{mod}\,8)$ and $3n^2 \equiv 3 \ (\mathrm{mod}\,8)$. Because $2a^2 \equiv 0$ or 2 $(\mathrm{mod}\,8)$ and $10m^2 \equiv 0$ or 2 $(\mathrm{mod}\,8)$ this leads to a contradiction. Hence b and n must be even and from the initial equation we deduce that a is also even. This contradicts the assumption that $\gcd(a, b, c, d) = 1$.

Problem 3.64 Prove that the system of equations

$$\begin{cases} x^2 + 6y^2 = z^2, \\ 6x^2 + y^2 = t^2 \end{cases}$$

has no positive integer solutions.

Solution As in the previous solution, we can assume that $\gcd(x, y, z, t) = 1$. Adding up the equations yields

$$7\left(x^2 + y^2\right) = z^2 + t^2.$$

The square residues modulo 7 are 0, 1, 2, and 4. It is not difficult to see that the only pair of residues which add up to 0 modulo 7 is $(0, 0)$, hence z and t are divisible by 7. Setting $z = 7z_1$ and $t = 7t_1$ yields

$$7\left(x^2 + y^2\right) = 49\left(z_1^2 + t_1^2\right)$$

or

$$x^2 + y^2 = 7\left(z_1^2 + t_1^2\right).$$

It follows that x and y are also divisible by 7, contradicting the fact that $\gcd(x, y, z, t) = 1$.

Problem 3.65 Let k and n be positive integers, with $n > 2$. Prove that the equation

$$x^n - y^n = 2^k$$

has no positive integer solutions.

Solution We may assume x and y are odd, otherwise dividing both x and y by 2 yields a smaller solution. If $n = 2m$ $(m \geq 2)$ is even, then we factor to get $(x^m - y^m)(x^m + y^m) = 2^k$. Hence both factors are powers of 2, say $x^m - y^m = 2^r$ and $x^m + y^m = 2^s$, with $s > r$. Solving gives $x^m = 2^{s-1} + 2^{r-1}$ and $y^m = 2^{s-1} - 2^{r-1}$. Since x^m and y^m are odd integers, this forces $r = 1$. But then x^m and y^m are two mth powers which differ by 2, a contradiction. If n is odd, then we factor as

$$(x - y)\left(x^{n-1} + x^{n-2}y + \cdots + y^{n-1}\right) = 2^k.$$

The second factor is odd since n is odd and each of the n terms is odd. Hence it must be 1, which is again a contradiction.

Problem 3.66 Prove that the equation

$$\frac{x^{2000} - 1}{x - 1} = y^2$$

has no positive integer solutions.

Solution Observe that

$$\frac{x^{2000} - 1}{x - 1} = \left(x^{1000} + 1\right)\left(x^{500} + 1\right)\frac{x^{500} - 1}{x - 1}.$$

Setting

$$a = x^{1000} + 1,$$
$$b = x^{500} + 1,$$
$$c = \frac{x^{500} - 1}{x - 1},$$

we see that b and c divide $a - 2$ and c divides $b - 2$. It follows that the greatest common divisor of any two of a, b, c is at most 2. The product abc is a square only if a, b, c are squares or doubles of squares. It is not difficult to see that a and b cannot be squares, hence they are doubles of squares. This implies that

$$4ab = 4x^{1500} + 4x^{1000} + 4x^{500} + 4$$

is a square. But this is impossible, since a short computation shows that

$$4x^{1500} + 4x^{1000} + 4x^{500} + 4 > \left(2x^{750} + x^{250}\right)^2$$

and

$$4x^{1500} + 4x^{1000} + 4x^{500} + 4 < \left(2x^{750} + x^{250} + 1\right)^2.$$

Problem 3.67 Prove that the equation

$$4\left(x_1^4 + x_2^4 + \cdots + x_{14}^4\right) = 7\left(x_1^3 + x_2^3 + \cdots + x_{14}^3\right)$$

has no solution in positive integers.

Solution Suppose, by way of contradiction, that such a solution exists. Then

$$\sum_{k=1}^{14}\left(x_k^4 - \frac{7}{4}x_k^3\right) = 0.$$

Observe that

$$\sum(x_k - 1)^4 = \sum\left(x_k^4 - 4x_k^3 + 6x_k^2 - 4x_k + 1\right)$$

$$= \sum\left(x_k^4 - \frac{7}{4}x_k^3 - \frac{9}{4}x_k^3 + 6x_k^2 - 4x_k + 1\right)$$

$$= \sum\left(-\frac{9}{4}x_k^3 + 6x_k^2 - 4x_k + 1\right).$$

Since

$$-\frac{9}{4}x_k^3 + 6x_k^2 - 4x_k = -x_k\left(\frac{3}{2}x_k - 2\right)^2 \le 0$$

we obtain

$$\sum(x_k - 1)^4 \le 14.$$

This inequality implies that each x_k is equal to either 1 or 2. Now, suppose x of the numbers $x_1, \ldots, x_{14}$ are equal to 1 and y of them are equal to 2. Then $x + y = 14$ and the original equation gives

$$4(x + 16y) = 7(x + 8y).$$

We obtain $x = \frac{112}{11}$ and $y = \frac{42}{11}$, a contradiction.

Problem 3.68 Prove that the equation

$$x^2 + y^2 + z^2 = 2011^{2011} + 2012$$

has no solution in integers.

Solution We begin by noticing that a square can only equal 0, 1, or 4 modulo 8. Let us check the right-hand side modulo 8. We have

$$2011 \equiv 3 \pmod 8,$$

hence

$$2011^{2011} \equiv 3^{2011} \equiv \left(3^2\right)^{1005} \cdot 3 \equiv 3 \pmod 8,$$

and since

$$2012 \equiv 4 \pmod 8,$$

it follows that

$$2011^{2011} + 2012 \equiv 7 \pmod 8.$$

Finally, observe that three (not necessarily distinct) numbers from the set $\{0, 1, 4\}$ cannot add to 7 (mod 8), so the equation has no solutions in integers.

Problem 3.69 Prove that the system

$$x^6 + x^3 + x^3 y + y = 147^{157},$$

$$x^3 + x^3 y + y^2 + y + z^9 = 157^{147}$$

has no solution in integers x, y, and z.

Solution Adding the equations yields

$$x^6 + y^2 + 2x^3 y + 2x^3 + 2y + z^9 = 147^{157} + 157^{147}.$$

The first five terms in the left-hand side suggest the expansion of a square, so add 1 to both sides to obtain

$$\left(x^3 + y + 1\right)^2 + z^9 = 147^{157} + 157^{147} + 1.$$

We will check both sides of the equation modulo 19. By Fermat's little theorem, if z is not divisible by 19, then $z^{18} \equiv 1 \pmod{19}$, and it follows that $z^9 \equiv \pm 1 \pmod{19}$. On the other hand, the possible remainders of a square modulo 19 are $0, 1, 4, 5, 6, 7, 9, 11, 16$, and 17. Therefore the left-hand side can take any value modulo 19 except for 13 and 14.

For the right-hand side, we have (all congruences are mod 19)

$$147^{157} \equiv 14^{157} \equiv \left(14^{18}\right)^8 \cdot 14^{13} \equiv 14^{13} \equiv (-5)^{13} \equiv -5^{13}$$

$$\equiv -5^{-5} \equiv -4^5 \equiv -1024 \equiv 2.$$

Similarly,

$$157^{147} \equiv 11 \pmod{19}.$$

It follows that the right-hand side is congruent to 14 modulo 19, hence the equation has no solutions in integers.

Problem 3.70 Prove that the equation

$$x^5 + y^5 + 1 = (x+2)^5 + (y-3)^5$$

has no solution in integers.

Solution Assume the contrary. Using Fermat's little theorem and taking both sides modulo 5 we obtain an obvious contradiction.

Observation Why did we choose $p = 19$ to check the equation mod p? Generally speaking, if in a Diophantine equation we have a term like x^k and $p = 2k + 1$ is a prime number, it is a good idea to check the equation mod p. That is because Fermat's little theorem gives $x^{2k} = (x^k)^2 \equiv 1 \pmod{p}$, if p does not divide x. Hence x^k can only take the values $-1, 0$, and $1 \pmod{p}$.

Problem 3.71 Prove that the equation

$$x^5 = y^2 + 4$$

has no solution in integers.

Solution Taking into account the observation from the previous solution, we will check the equation mod 11. The left-hand side can be $-1, 0$, or 1. The possible remainders of a square mod 11 are $0, 1, 3, 4, 5, 9$, hence the right-hand side can be $4, 5, 7, 8, 9$, or 2. Thus, the equality is never possible mod 11.

Problem 3.72 Prove that the equation

$$x^3 - 3xy^2 + y^3 = 2891$$

has no solution in integers.

Solution Observe that $2891 = 7^2 \cdot 59$. If one of x, y is divisible by 7, then so is the other one, and it follows that 7^3 divides 2891, a contradiction. Since $7 \nmid y$, there exists z such that $yz \equiv 1 \pmod 7$. Multiplying by z^3 and denoting $xz = t$, we obtain

$$t^3 - 3t + 1 \equiv 0 \pmod 7.$$

6.7 Powers of 2

Problem 3.75 Let n be a positive integer such that $2^n + 1$ is a prime number. Prove that $n = 2^k$, for some integer k.

Solution Suppose that n is not a power of 2. Then it can be written in the form $n = 2^k(2p + 1)$, with $k \geq 0$ and $p \geq 1$. But then we have

$$2^n + 1 = \left(2^{2^k}\right)^{2p+1} + 1 = \left(2^{2^k} + 1\right)\left(\left(2^{2^k}\right)^{2p} - \left(2^{2^k}\right)^{2p-1} + \cdots - 2^{2^k} + 1\right),$$

hence $2^n + 1$ is not a prime number.

Observation The numbers $F_n = 2^{2^n} + 1$ are called Fermat's numbers. Fermat conjectured that all such numbers are prime, which is true for $n \leq 4$, but Euler proved that $F_5 = 2^{32} + 1$ is divisible by 641. The proof is quite short: notice that $641 = 640 + 1 = 5 \cdot 2^7 + 1$, hence $5 \cdot 2^7 \equiv -1 \pmod{641}$, so that $5^4 \cdot 2^{28} \equiv 1 \pmod{641}$. On the other hand, $641 = 625 + 16 = 5^4 + 2^4$, so $2^4 \equiv -5^4 \pmod{641}$. Multiplying these two congruencies, we obtain $5^4 \cdot 2^{32} \equiv -5^4 \pmod{641}$, hence $2^{32} \equiv -1 \pmod{641}$.

Problem 3.76 Let n be a positive integer such that $2^n - 1$ is a prime number. Prove that n is a prime number.

Solution If n is not a prime number, then $n = ab$, for some positive integers $a, b > 1$. We obtain

$$2^n - 1 = 2^{ab} - 1 = \left(2^a\right)^b - 1 = \left(2^a - 1\right)\left(\left(2^a\right)^{b-1} + \left(2^a\right)^{b-2} + \cdots + 2^a + 1\right),$$

and this factorization shows that $2^n - 1$ is not a prime number.

Problem 3.77 Prove that the number $A = 2^{1992} - 1$ can be written as a product of 6 integers greater than 2^{248}.

Solution It is again an exercise in factorization:

$$A = 2^{1992} - 1 = 2^{249 \cdot 8} - 1 = \left(2^{249}\right)^8 - 1$$

$$= \left(2^{249} - 1\right)\left(2^{249} + 1\right)\left(\left(2^{249}\right)^2 + 1\right)\left(\left(2^{249}\right)^4 + 1\right).$$

Now, observe that

$$\left(2^{249}\right)^2 + 1 = \left(2^{249}\right)^2 + 2 \cdot 2^{249} + 1 - 2^{250}$$

$$= \left(2^{249} + 1\right)^2 - \left(2^{125}\right)^2 = \left(2^{249} - 2^{125} + 1\right)\left(2^{249} + 2^{125} + 1\right),$$

and

$$\left(2^{249}\right)^4 + 1 = 2^{996} + 1 = \left(2^{332}\right)^3 + 1$$

$$= \left(2^{332} + 1\right)\left(2^{664} - 2^{332} + 1\right).$$

Thus, the six integers are $2^{249} - 1, 2^{249} + 1, 2^{249} - 2^{125} + 1, 2^{249} + 2^{125} + 1, 2^{332} + 1$ and $2^{664} - 2^{332} + 1$. It is not difficult to see that all six are greater than 2^{248}.

Problem 3.78 Determine the remainder of $3^{2^n} - 1$ when divided by 2^{n+3}.

Solution Observe that after we use n times the identity

$$x^2 - 1 = (x - 1)(x + 1),$$

we obtain

$$3^{2^n} - 1 = (3 - 1)(3 + 1)\left(3^2 + 1\right)\left(3^{2^2} + 1\right)\cdots\left(3^{2^{n-1}} + 1\right).$$

Now, each of the numbers $3^2 + 1, 3^{2^2} + 1, \ldots, 3^{2^{n-1}} + 1$ is divisible by 2 but not by 4. Indeed, $3 \equiv -1 \pmod 4$, so $3^{2^k} \equiv (-1)^{2^k} \equiv 1 \pmod 4$ and it follows that $3^{2^k} + 1 \equiv 2 \pmod 4$. We deduce that there exists an odd integer $2m + 1$ such that

$$\left(3^2 + 1\right)\left(3^{2^2} + 1\right)\cdots\left(3^{2^{n-1}} + 1\right) = 2^{n-1}(2m + 1).$$

Then

$$3^{2^n} - 1 = 2 \cdot 4 \cdot 2^{n-1}(2m + 1) = m \cdot 2^{n+3} + 2^{n+2},$$

and this shows that the requested remainder is 2^{n+2}.

Problem 3.79 Prove that for each n, there exists a number A_n, divisible by 2^n, whose decimal representation contains n digits, each of them equal to 1 or 2.

Solution We prove the assertion by induction on n. For $n = 1$, take $A_1 = 2$; for $n = 2$, take $A_2 = 12$. Now, suppose there exists A_n, divisible by 2^n, whose decimal representation contains n digits, each of them equal to 1 or 2.

If 2^{n+1} divides A_n, then A_{n+1} is obtained by adding the digit 2 at the beginning of A_n. Thus,

$$A_{n+1} = 2 \cdot 10^n + A_n = 2^{n+1} \cdot 5^n + A_n,$$

and we see that this number has $n + 1$ digits 1 or 2 and it is divisible by 2^{n+1}.

If 2^{n+1} does not divide A_n, then $A_n = 2^n k$ for some odd integer k. In this case, we add the digit 1 at the beginning of A_n. It follows that

$$A_{n+1} = 10^n + A_n = 2^n \cdot 5^n + 2^n \cdot k = 2^n\left(5^n + k\right).$$

The claim is proved by observing that since k is odd, $5^n + k$ is even, thus A_{n+1} is divisible by 2^{n+1}.

Observation We can prove that the number A_n is unique. Indeed, suppose B_n is another number with the enounced properties. Then $A_n - B_n$ is divisible by 2^n, hence they have the same last digit. Let A_{n-1} and B_{n-1} be the numbers obtained by discarding the last digit of A_n and B_n. Then $A_n - B_n = 10(A_{n-1} - B_{n-1})$, and we deduce that $A_{n-1} - B_{n-1}$ is divisible by 2^{n-1}, hence, again, they have the same last digit. Repeating this argument, we finally deduce that $A_n = B_n$. We can see that we did not use the fact that A_n and B_n are divisible by 2^n, but only that their difference is divisible by 2^n. This shows that the 2^n numbers with n digits equal to 1 or 2 give different remainders when divided by 2^n. Finally, we observe that the digits 1 and 2 from the enounce can be replaced with any two other non-zero digits with different parities.

Problem 3.80 Using only the digits 1 and 2, one writes down numbers with 2^n digits such that the digits of every two of them differ in at least 2^{n-1} places. Prove that no more than 2^{n+1} such numbers exist.

Solution Let us suppose that $2^{n+1} + 1$ such numbers exist. From these, at least $2^n + 1$ have the same last digit. Denote by A the set of these numbers. For $a, b \in A$, let $c(a, b)$ be the number of digit coincidences. Thus, from the enounce it follows that $c(a, b) \le 2^{n-1}$, for every a and b. Let N the total number of coincidences for all numbers in A. Then

$$N \le 2^{n-1} \cdot \binom{2^n + 1}{2} = 2^{3n-2} + 2^{2n-2}.$$

On the other hand, if we arrange the numbers of A in a table with $2^n + 1$ rows and 2^n columns, we count in the last column $\binom{2^n+1}{2}$ coincidences and in each of the remaining $2^n - 1$ columns at least $2^{2(n-1)}$ coincidences. Indeed, if on some column we have $2^{n-1} - k$ digits of one type and $2^{n-1} + k + 1$ digits of the other type, then the number of coincidences equals

$$\binom{2^{n-1} - k}{2} + \binom{2^{n-1} + k + 1}{2} = 2^{2(n-1)} + k^2 + k \ge 2^{2(n-1)}.$$

We conclude that

$$N \ge \binom{2^n + 1}{2} + (2^n - 1) \cdot 2^{2(n-1)} = 2^{3n-2} + 2^{2n-2} + 2^{n-1}.$$

which is a contradiction.

Observation It can be shown that 2^{n+1} numbers with the required properties exist. We prove this inductively. For $n = 1$, take the numbers 11, 12, 21 and 22, which clearly satisfy the conditions. Suppose there exists a set S_n with 2^{n+1} numbers having 2^n digits such that every two of them differ in at least 2^{n-1} places. We construct a set S_{n+1} with 2^{n+2} numbers having 2^{n+1} digits generating from each element of S_n two elements of S_{n+1} as follows: if $a = \overline{a_1 a_2 \ldots a_n} \in S_n$, then in S_{n+1} we put $b = \overline{a_1 a_2 \ldots a_n a_1 a_2 \ldots a_n}$ and $c = \overline{a_1 a_2 \ldots a_n a'_1 a'_2 \ldots a'_n}$, where $a'_k = 1$ if $a_k = 2$ and $a'_k = 2$ if $a_k = 1$. For instance, if $S_2 = \{11, 12, 21, 22\}$, then $S_3 = \{1111, 1212, 2121, 2222, 1122, 1221, 2112, 2211\}$. It is not difficult to check that the set thus obtained has the required properties.

Problem 3.81 Does there exist a natural number N which is a power of 2 whose digits (in the decimal representation) can be permuted to form a different power of 2?

Solution The answer is negative. Indeed, suppose such N exists and let N' another power of 2 obtained by rearranging the digits of N. We can assume that $N < N'$. Since N and N' have the same number of digits, we must have $N' = 2N$, or $N' = 4N$ or $N' = 8N$. It results that

$$N' - N \in \{N, 3N, 7N\}.$$

We get a contradiction from the following.

Lemma Let $n = \overline{a_k a_{k-1} \ldots a_1 a_0}$ be a positive integer and

$$s(n) = a_k + a_{k-1} + \cdots + a_1 + a_0$$

the sum of its digits. Then the difference $n - s(n)$ is divisible by 9.

Proof We have

$$n - s(n) = a_k \cdot 10^k + a_{k-1} \cdot 10^{k-1} + \cdots + a_1 \cdot 10 + a_0 - a_k - a_{k-1} - \cdots$$
$$- a_1 - a_0$$
$$= a_k(10^k - 1) + a_{k-1}(10^{k-1} - 1) + \cdots + a_1(10 - 1),$$

and the conclusion follows by noticing that every number of the form $10^p - 1$ is divisible by 9.

Returning to our problem, we see that since N and N' have the same digits, $s(N) = s(N')$, thus $N' - N$ must be divisible by 9. This is impossible, because $N' - N \in \{N, 3N, 7N\}$ and none of the numbers $N, 3N, 7N$ is divisible by 9. $\square$

Problem 3.82 For a positive integer N, let $s(N)$ the sum of its digits, in the decimal representation. Prove that there are infinitely many n for which $s(2^n) > s(2^{n+1})$.

Solution We use again the above lemma. Suppose, by way of contradiction, that there exist finitely many n for which $s(2^n) > s(2^{n+1})$. Then there exists m such that for every $n > m$, $s(2^n) \le s(2^{n+1})$. Since $s(2^n) = s(2^{n+1})$ if and only if $2^{n+1} - 2^n = 2^n$ is divisible by 9, we deduce that the sequence $s(2^n)$ is strictly increasing for $n > m$. Moreover, since the remainders of the numbers 2^n divided by 9 are $2, 4, 8, 7, 5, 1$ and repeat periodically after 6 steps, we deduce that

$$s(2^{n+6}) \ge s(2^n) + 2 + 4 + 8 + 7 + 5 + 1 = s(2^n) + 27$$

and then, inductively, that

$$s(2^{n+6k}) \ge s(2^n) + 27k,$$

for $n > m$ and all positive integers k.

On the other hand, $2^3 < 10$, so $2^{n+6k} < 10^{\frac{n}{3}+2k}$ and this shows that the number 2^{n+6k} has at most $\frac{n}{3} + 2k$ digits in its decimal representation. It follows that

$$s\left(2^{n+6k}\right) \leq 9\left(\frac{n}{3} + 2k\right) = 3n + 18k.$$

Combining the two estimates of $s(2^{n+6k})$, we deduce

$$s\left(2^n\right) + 27k \leq 3n + 18k,$$

for every positive integer k. This is a contradiction.

Problem 3.83 Find all integers of the form 2^n (where n is a natural number) such that after deleting the first digit of its decimal representation we again get a power of 2.

Solution Suppose 2^m is obtained after deleting the first digit (equal to a) of the decimal representation of 2^n. We have then $2^n = 10^k a + 2^m$, for some integer k, hence $2^{n-m} - 1$ is divisible by 5. Checking the remainders of the powers of 2 divided by 5, it results that $2^{n-m} - 1$ is divisible by 5 if and only if $n - m = 4t$, for some positive integer t. But then

$$10^k a = 2^m \left(2^{4t} - 1\right) = 2^m \left(2^{2t} + 1\right)\left(2^{2t} - 1\right) = 2^m \left(2^{2t} + 1\right)\left(2^t + 1\right)\left(2^t - 1\right).$$

Observe that $2^{2t} + 1$ and $2^{2t} - 1$ are odd integers differing by 2, therefore are relatively prime. The same applies to $2^t + 1$ and $2^t - 1$, hence $2^{2t} + 1, 2^t + 1$ and $2^t - 1$ are all odd, pairwise prime integers. If $t > 1$, then each of the three numbers has an odd prime divisor, but this is impossible, since $10^k a$ is divisible by 5 and at most one of the numbers 3 and 7 (if any of them divides a). Consequently, $t = 1$, hence $10^k a = 2^m \cdot 3 \cdot 5$, which leads to $k = 1$ and $a = 2^{m-1} \cdot 3$. The only possible cases are $m = 1$ and $m = 2$, so the solutions to the problem are $2^5 = 32$ and $2^6 = 64$.

Problem 3.84 Let $a_0 = 0, a_1 = 1$ and, for $n \geq 2, a_n = 2a_{n-1} + a_{n-2}$. Prove that a_n is divisible by 2^k if and only if n is divisible by 2^k.

Solution Using the standard algorithm for recurrence relations, we obtain

$$a_n = \frac{1}{2\sqrt{2}}[(1 + \sqrt{2})^n - (1 - \sqrt{2})^n],$$

for every positive integer n. If we define

$$b_n = \frac{1}{2}[(1 + \sqrt{2})^n + (1 - \sqrt{2})^n],$$

we observe that b_n satisfies the same recurrence relation, and since $b_0 = b_1 = 1$, all b_n are positive integers.

We have

$$b_n + a_n \sqrt{2} = \left(1 + \sqrt{2}\right)^n$$

and

$$b_n - a_n \sqrt{2} = \left(1 - \sqrt{2}\right)^n.$$

Multiplying these equalities we obtain $b_n^2 - 2a_n^2 = (-1)^n$, hence b_n is odd for each n.

Now, we prove the claim by induction on k. For $k = 0$ we have to prove that a_n is odd if and only if n is odd. We have

$$2a_n^2 = b_n^2 - (-1)^n$$

and $b_n = 2m + 1$, for some integer m, hence

$$a_n^2 = 2(m^2 + m) + \frac{1 - (-1)^n}{2}.$$

It follows that a_n is odd if and only if $1 - (-1)^n = 2$, that is, n is odd. The inductive step follows from the equality

$$a_{2n} = 2a_n b_n,$$

which is obtained observing that

$$b_{2n} + a_{2n} \sqrt{2} = \left(1 + \sqrt{2}\right)^{2n} = \left(b_n + a_n \sqrt{2}\right)^2 = b_n^2 + 2a_n^2 + 2a_n b_n \sqrt{2}.$$

Problem 3.85 If $A = \{a_1, a_2, \ldots, a_p\}$ is a set of real numbers such that $a_1 > a_2 > \cdots > a_p$, we define

$$s(A) = \sum_{k=1}^{p} (-1)^{k-1} a_k.$$

Let M be a set of n positive integers. Prove that $\sum_{A \subseteq M} s(A)$ is divisible by 2^{n-1}.

Solution Let $M = \{a_1, a_2, \ldots, a_n\}$. We can assume that

$$a_1 > a_2 > \cdots > a_n.$$

Let $A \subseteq M - \{a_1\}$, $A = \{a_{i_1}, a_{i_2}, \ldots, a_{i_s}\}$, with $a_{i_1} > a_{i_2} > \cdots > a_{i_s}$, and $A' = A \cup \{a_1\}$. Then

$$a_1 > a_{i_1} > a_{i_2} > \cdots > a_{i_s}$$

and we have

$$s(A) + s(A') = a_{i_1} - a_{i_2} + \cdots + (-1)^{s-1} a_{i_s} + a_1 - a_{i_1} + a_{i_2} - \cdots + (-1)^s a_{i_s} = a_1.$$

Now, matching all the 2^{n-1} subsets A of the set $M - \{a_1\}$ with the corresponding subsets $A' = A \cup \{a_1\}$, we find

$$\sum_{A \subseteq M} s(A) = 2^{n-1} a_1,$$

and the claim is proved.

Problem 3.86 Find all positive integers a, b, such that the product

$$(a + b^2)(b + a^2)$$

is a power of 2.

Solution We begin by noticing that a and b must have the same parity. Let $a + b^2 = 2^m$ and $b + a^2 = 2^n$, and assume that $m \geq n$. Then $a + b^2 \geq b + a^2$, or $b^2 - b \geq a^2 - a$, which implies $b \geq a$. Subtracting the two equalities yields

$$(b - a)(a + b - 1) = 2^m - 2^n = 2^n (2^{m-n} - 1).$$

Because $a + b - 1$ is odd, it follows that 2^n divides $b - a$. Let $b - a = 2^n c$, for some positive integer c. Then $b = 2^n c + a = 2^n - a^2$, therefore $a + a^2 = 2^n (1 - c)$, which implies $c = 0$, hence $a = b$.

Finally, we deduce that $a(a + 1)$ is a power of 2, and this is possible only for $a = 1$.

Problem 3.87 Let $f(x) = 4^x + 6^x + 9^x$. Prove that if m and n are positive integers, then $f(2^m)$ divides $f(2^n)$ whenever $m \leq n$.

Solution Define $g(x) = 4^x - 6^x + 9^x$. From the identity

$$(a^2 + ab + b^2)(a^2 - ab + b^2) = a^4 + a^2 b^2 + b^4$$

taking $a = 2$ and $b = 3$, we deduce that

$$f(x)g(x) = f(2x).$$

Iterating this, we get

$$f(x)g(x)g(2x) \cdots g(2^{k-1} x) = f(2^k x),$$

for all $k \geq 2$.

Now, suppose $m \leq n$. Then

$$f(2^m)g(2^m)g(2^{m+1}) \cdots g(2^{n-1}) = f(2^n),$$

hence $f(2^m)$ divides $f(2^n)$.

Problem 3.88 Show that, for any fixed integer $n \geq 1$, the sequence

$$2, 2^2, 2^{2^2}, 2^{2^{2^2}}, \ldots \pmod{n}$$

is eventually constant.

Solution For a more comfortable notation, define $x_0 = 1$ and $x_{m+1} = 2^{x_m}$, for $m \geq 0$. We have to prove that the sequence $(x_m)_{m \geq 0}$ is eventually constant mod n.

We will prove by induction on n the following statement: $x_k \equiv x_{k+1} \pmod{n}$, for all $k \geq n$.

For $n = 1$ there is nothing to prove. Assume that the statement is true for all numbers less than n. If $n = 2^a \cdot b$, with odd b, it suffices to prove that $x_k \equiv x_{k+1} \pmod{2^a}$, and $x_k \equiv x_{k+1} \pmod{b}$, for all $k \geq n$.

It is not difficult to prove inductively that $x_n > n$. Therefore, if $k \geq n$, we have $x_k \equiv 0 \pmod{2^a}$ (x_k is a power of 2 greater than n, thus greater than 2^a), whence $x_k \equiv x_{k+1} \pmod{2^a}$.

For the second congruence, observe that $x_k \equiv x_{k+1} \pmod{b}$ is equivalent to $2^{x_{k-1}} \equiv 2^{x_k} \pmod{b}$, or $2^{x_k - x_{k-1}} \equiv 1 \pmod{b}$. The latter is true whenever $x_{k-1} \equiv x_k \pmod{\phi(b)}$ (ϕ is Euler's totient function), but this follows from the induction hypothesis, since $\phi(b) \leq b - 1 \leq n - 1$.

6.8   Progressions

Problem 3.92 Partition the set of positive integers into two subsets such that neither of them contains a non-constant arithmetical progression.

Solution Partition the set $\{1, 2, 3, \ldots\}$ in the following way:

1		4	5	6						$\cdots$
	2	3			7	8	9	10	$\cdots$	

None of the sets

$$A = \{1, 4, 5, 6, 11, 12, 13, 14, 15, \ldots\}$$

or

$$B = \{2, 3, 7, 8, 9, 10, 16, 17, 18, 19, 20, 21, 22, \ldots\}$$

contains a non-constant arithmetical progression. Indeed, if such a progression is contained in one of the sets, let r be its common difference. But in both sets we can find a "gap" of more that r consecutive integers, a contradiction.

Problem 3.93 Prove that among the terms of the progression $3, 7, 11, \ldots$ there are infinitely many prime numbers.

Solution Suppose the contrary and let p be the greatest prime number in the given progression. Consider the number

$$n = 4p! - 1.$$

It is not difficult to see that it is a term of the progression (in fact, the given progression contains all positive integers of the form $4k - 1$). Thus n must be composite, since $n > p$. Observe that n is not divisible by any prime of the form $4k - 1$ (all these are factors in $p!$), hence all the prime factors of n are of the form $4k + 1$. The product of several factors of the form $4k + 1$ is again of the form $4k + 1$, hence $n = 4k + 1$, for some k. This is a contradiction.

Problem 3.94 Does there exist an (infinite) non-constant arithmetical progression whose terms are all prime numbers?

Solution The answer is negative. Indeed, if such progression exists, denote its common difference by r, and consider the consecutive r integers

$$(r + 1)! + 2, (r + 1)! + 3, \ldots, (r + 1)! + (r + 1).$$

Each of them is a composite number, but since the progression has the common difference r, one out of any r consecutive integers must be a term of the progression. This is a contradiction.

Problem 3.95 Consider an arithmetical progression of positive integers. Prove that one can find infinitely many terms the sum of whose decimal digits is the same.

Solution Let $(a_n)_{n \geq 1}$ be the progression and denote by r its common difference. Suppose the number a_1 has d digits (in its decimal representation). Then for all $k > d$ the digits sum of the number

$$a_1 + 10^k r$$

is the same.

Problem 3.96 The set of positive integers is partitioned into n arithmetical progressions, with common differences $r_1, r_2, \ldots, r_n$. Prove that

$$\frac{1}{r_1} + \frac{1}{r_2} + \cdots + \frac{1}{r_n} = 1.$$

Solution Let $a_k = a_1 + (k - 1)r_1$ the progression with common difference r_1. Let us count how many terms of this sequence are less than or equal to some positive integer N. The inequality $a_k \leq N$ is equivalent to

$$a_1 + (k - 1)r_1 \leq N$$

or

$$k \le \frac{N}{r_1} - \frac{a}{r_1} + 1.$$

It follows that the number of terms of the first progression belonging to the set $\{1, 2, \ldots, N\}$ equals

$$\left\lfloor \frac{N}{r_1} - \frac{a}{r_1} + 1 \right\rfloor.$$

Similarly, we deduce that the number of terms of the progression with common difference r_i belonging to the set $\{1, 2, \ldots, N\}$ equals

$$\left\lfloor \frac{N}{r_i} - \frac{a}{r_i} + 1 \right\rfloor.$$

Since the progressions form a partition of the set of positive integers, we must have

$$\sum_{i=1}^{n} \left\lfloor \frac{N}{r_i} - \frac{a}{r_i} + 1 \right\rfloor = N.$$

Using the inequality $\lfloor x \rfloor \le x < \lfloor x \rfloor + 1$, we obtain

$$N \le \sum_{i=1}^{n} \left(\frac{N}{r_i} - \frac{a}{r_i} + 1 \right) < N + n$$

hence

$$1 \le \sum_{i=1}^{n} \frac{1}{r_i} - \frac{1}{N} \sum_{i=1}^{n} \frac{na}{r_1} + \frac{n}{N} < 1 + \frac{n}{N}$$

and letting $N \to \infty$ yields the desired result.

Problem 3.97 Prove that for every positive integer n one can find n integers in arithmetical progression, all of them nontrivial powers of some integers, but one cannot find an infinite sequence with this property.

Solution We prove the assertion by induction on n. For $n = 3$, we can consider the numbers $1, 25, 49$. Suppose the assertion is true for some n and let $a_i = b_i^{k_i}, i = 1, 2, \ldots, n$, be the terms of the progression having the common difference d. Let $b = a_n + d$ and let $k = \mathrm{lcm}(k_1, k_2, \ldots, k_n)$. Then the $n + 1$ numbers

$$a_1 b^k, \ a_2 b^k, \ldots, \ a_n b^k, \ b^{k+1}$$

are in progression and all are power of some integers. Indeed, for all i, $1 \le i \le n$, there exists d_i such that $k = k_i d_i$ and we obtain

$$a_i b^k = b_i^{k_i} b^k = b_i^{k_i} b^{k_i d_i} = \left(b_i b^{d_i} \right)^{k_i},$$

as desired.

For the second part, we need a result from Calculus: if $a, b > 0$ then

$$\lim_{n \to \infty} \sum_{k=1}^{n} \frac{1}{ak + b} = +\infty.$$

From this we deduce that if $(a_n)_{n \geq 1}$ is a progression of positive integers, then

$$\lim_{n \to \infty} \sum_{k=1}^{n} \frac{1}{a_k} = +\infty.$$

Now suppose that $(a_n)_{n \geq 1}$ is a progression in which all terms are powers of some integers. Let S be the set of all positive integers greater than 1 that are powers of some integers. We will prove that

$$\sum_{a \in S} \frac{1}{a} \leq 1$$

which contradicts the previous result. Indeed, we have

$$\sum_{a \in S} \frac{1}{a} \leq \sum_{n \geq 2} \sum_{k \geq 2} \frac{1}{n^k} = \sum_{n \geq 2} \frac{1}{n^2} \left(1 + \frac{1}{n} + \frac{1}{n^2} + \cdots \right)$$

$$= \sum_{n \geq 2} \frac{1}{n^2} \cdot \frac{1}{1 - \frac{1}{n}} = \sum_{n \geq 2} \frac{1}{n(n-1)} = \sum_{n \geq 2} \left(\frac{1}{n-1} - \frac{1}{n} \right) = 1.$$

Problem 3.98 Prove that for any integer n, $n \geq 3$, there exist n positive integers in arithmetical progression $a_1, a_2, \ldots, a_n$ and n positive integers in geometric progression $b_1, b_2, \ldots, b_n$, such that

$$b_1 < a_1 < b_2 < a_2 < \cdots < b_n < a_n.$$

Solution Let $m > n^2$ be an integer. We define

$$B_k = \left(1 + \frac{1}{m} \right)^k,$$

for $k = 1, 2, \ldots, n$. Observe that for $k \geq 2$

$$B_k > 1 + \frac{k}{m}.$$

For $k \leq n$ we have

$$B_k = 1 + \frac{k}{m} + \frac{k(k-1)}{2!m^2} + \cdots + \frac{k!}{k!m^k}$$

$$\leq 1 + \frac{k}{m} + \frac{n(n-1)}{2!m^2} + \cdots + \frac{n(n-1)\cdots(n-k+1)}{k!m^k}$$

$$= 1 + \frac{k}{m} + \frac{1}{m}\left(\frac{n(n-1)}{2!m} + \cdots + \frac{n(n-1)\cdots(n-k+1)}{k!m^{k-1}}\right)$$

$$< 1 + \frac{k}{m} + \frac{1}{m}\left(\frac{1}{2!} + \frac{1}{3!} + \cdots + \frac{1}{k!}\right) < 1 + \frac{k+1}{m}.$$

We have used the fact that $m > n^2$ and the inequalities

$$\frac{1}{2!} + \frac{1}{3!} + \cdots + \frac{1}{k!} < \frac{1}{1\cdot 2} + \frac{1}{2\cdot 3} + \cdots + \frac{1}{(k-1)\cdot k} = 1 - \frac{1}{k} < 1.$$

Defining

$$A_k = 1 + \frac{k+1}{n}$$

for $1 \le k \le n$, yields

$$B_1 < A_1 < B_2 < A_2 < \cdots < B_n < A_n.$$

In order to obtain progressions with integer terms we define $a_k = nm^n A_k$ and $b_k = nm^n B_k$, for all k, $1 \le k \le n$.

Problem 3.99 Let $(a_n)_{n\ge 1}$ be an arithmetic sequence such that a_1^2, a_2^2, and a_3^2 are also terms of the sequence. Prove that the terms of this sequence are all integers.

Solution Let d be the common difference. If $d = 0$, then it is not difficult to see that either $a_n = 0$ for all n, or $a_n = 1$ for all n. Suppose that $d \ne 0$, and consider the positive integers m, n, p, such that $a_1^2 = a_1 + md$, $(a_1 + d)^2 = a_1 + nd$, and $(a_1 + 2d)^2 = a_1 + pd$. Subtracting the first equation from the other two yields

$$\begin{cases} 2a_1 + d = n - m, \\ 4a_1 + 4d = p - m \end{cases}$$

and solving for a_1 and d we obtain

$$a_1 = \frac{1}{4}(4n - 3m - p),$$

$$d = \frac{1}{2}(m - 2n + p),$$

hence both a_1 and d are rational numbers.

Observe that the equation $a_1^2 = a_1 + md$ can be written as

$$a_1^2 + (2m - 1)a_1 - m(d + 2a_1) = 0,$$

or

$$a_1^2 + (2m - 1)a_1 - m(n - m) = 0,$$

hence a_1 is the root of the polynomial with integer coefficients

$$P(x) = x^2 + (2m - 1)x - m(n - m).$$

By the integer root theorem it follows that a_1 is an integer, hence $d = n - m - 2a_1$ is an integer as well. We conclude that all terms of the sequence are integer numbers.

Problem 3.100 Let $A = \{1, \frac{1}{2}, \frac{1}{3}, \frac{1}{4}, \ldots\}$. Prove that for every positive integer $n \geq 3$ the set A contains a non-constant arithmetic sequence of length n, but it does not contain an infinite non-constant arithmetic sequence.

Solution For $n = 3$ we have the almost obvious example

$$\frac{1}{6}, \frac{1}{3}, \frac{1}{2}.$$

Writing this as

$$\frac{1}{6}, \frac{2}{6}, \frac{3}{6}$$

might give us a clue for the general case. Indeed, consider the arithmetic sequence

$$\frac{1}{n!}, \frac{2}{n!}, \ldots, \frac{n}{n!}.$$

When we write the fractions in their lowest terms, we see that all belong to A.

For the second part, just observe that every non-constant, infinite arithmetic progression is necessarily an unbounded sequence. Since A is bounded, it cannot contain such a sequence.

Problem 3.101 Let n be a positive integer and let $x_1 \leq x_2 \leq \cdots \leq x_n$ be real numbers. Prove that

$$\left(\sum_{i,j=1}^{n} |x_i - x_j| \right)^2 \leq \frac{2(n^2 - 1)}{3} \sum_{i,j=1}^{n} (x_i - x_j)^2.$$

Show that the equality holds if and only if $x_1, \ldots, x_n$ is an arithmetic sequence.

Solution Suppose the given number are the terms of an arithmetic sequence, with common difference d. Then $x_i - x_j = (i - j)d$ and the equality we have to prove becomes

$$\left(\sum_{i,j=1}^{n} |i - j| \right)^2 = \frac{2(n^2 - 1)}{3} \sum_{i,j=1}^{n} (i - j)^2.$$

We have

$$\sum_{i,j=1}^{n} |i-j| = \sum_{i=1}^{n}\sum_{j=1}^{n} |i-j| = \sum_{i=1}^{n}\left(\sum_{j=1}^{i}(i-j) + \sum_{j=i+1}^{n}(j-i)\right)$$

$$= \sum_{i=1}^{n}\left(i^2 - \frac{i(i+1)}{2} + \frac{n(n+1)}{2} - \frac{i(i+1)}{2} - i(n-i)\right)$$

$$= \sum_{i=1}^{n}\left(i^2 - (n+1)i + \frac{n(n+1)}{2}\right)$$

$$= \frac{n(n+1)(2n+1)}{6} - \frac{n(n+1)^2}{2} + \frac{n^2(n+1)}{2}$$

$$= \frac{n(n^2-1)}{3}.$$

On the other hand,

$$\sum_{i,j=1}^{n} (i-j)^2 = \sum_{i=1}^{n}\left(\sum_{j=1}^{n}(i^2 - 2ij + j^2)\right)$$

$$= \sum_{i=1}^{n}\left(ni^2 - in(n+1) + \frac{n(n+1)(2n+1)}{6}\right)$$

$$= \frac{n^2(n+1)(2n+1)}{6} - \frac{n^2(n+1)^2}{2} + \frac{n^2(n+1)(2n+1)}{6}$$

$$= \frac{n^2(n^2-1)}{6}.$$

Thus, the equality to prove is written as

$$\left(\frac{n(n^2-1)}{3}\right)^2 = \frac{2(n^2-1)}{3} \cdot \frac{n^2(n^2-1)}{6},$$

obviously true.

In order to prove the inequality, observe that

$$\sum_{i,j=1}^{n} (x_i - x_j)^2 = \sum_{i=1}^{n}\sum_{j=1}^{n}(x_i^2 - 2x_ix_j + x_j^2)$$

$$= \sum_{i=1}^{n}\left(nx_i^2 - 2x_i\sum_{j=1}^{n}x_j + \sum_{j=1}^{n}x_j^2\right)$$

$$= 2n\sum_{i=1}^{n}x_i^2 - 2\left(\sum_{i=1}^{n}x_i\right)^2.$$

In a similar way and taking into account the ordering of the x_i's, we obtain

$$\sum_{i,j=1}^{n} |x_i - x_j| = 2\sum_{i=1}^{n} (2i - n - 1)x_i.$$

Thus, the inequality becomes

$$\left(\sum_{i=1}^{n} (2i - n - 1)x_i\right)^2 \le \frac{n^2 - 1}{3}\left(n\sum_{i=1}^{n} x_i^2 - \left(\sum_{i=1}^{n} x_i\right)^2\right).$$

The key observation is that we can replace all x_i's by $x_i + c$, for an arbitrary constant c. Indeed, if we look at the original inequality, we see that the differences $x_i - x_j$ remain unchanged. Therefore, we can choose the constant c such that $\sum_{i=1}^{n} x_i = 0$. In this way we only have to check that

$$\left(\sum_{i=1}^{n} (2i - n - 1)x_i\right)^2 \le \frac{n^2 - 1}{3}\left(n\sum_{i=1}^{n} x_i^2\right),$$

which follows easily from Cauchy–Schwarz if we observe that

$$\sum_{i=1}^{n} (2i - n - 1) = \frac{n(n^2 - 1)}{3}.$$

The equality occurs if there exists d such that $x_i = d(2i - n - 1)$, for all i, hence if the numbers $x_1, x_2, \ldots, x_n$ form an arithmetic sequence.

6.9 The Marriage Lemma

Problem 3.104 A deck of cards is arranged, face up, in a 4×13 array. Prove that one can pick a card from each column in such a way as to get one card of each denomination.

Solution Consider the columns as boys and the denominations as girls. A boy is acquainted with a girl if in that column there exists a card of the respective denomination. Now choose k boys. They are acquainted with $4k$ (not necessarily distinct) girls. But each girl appear at most four times, since there are four cards of each denomination. Therefore, the number of distinct girls acquainted to the k boys is at least k, hence Hall's condition holds. The matching between the columns and the denominations show us how to pick the cards.

Problem 3.105 An $n \times n$ table is filled with 0 and 1 so that if we chose randomly n cells (no two of them in the same row or column) then at least one contains 1. Prove that we can find i rows and j columns so that $i + j \ge n + 1$ and their intersection contains only 1's.

Solution Let the rows be the boys and the columns be the girls. A boy is acquainted with a girl if at the intersection of the respective row and column there is a 0. Then the hypothesis simply says that there is no matching between the boys and the girls.

We deduce that Hall's condition is violated, hence we can find i rows such that such that the columns they are acquainted with are at most $i - 1$. But then at the intersections of these i rows and the remaining $j \geq n - i + 1$ columns there are only 1's. The conclusion is obvious.

Problem 3.106 Let X be a finite set and let $\bigsqcup_{i=1}^{n} X_i = \bigsqcup_{j=1}^{n} Y_j$ be two disjoint decompositions with all sets X_i's and Y_j's having the same size. Prove that there exist distinct elements $x_1, x_2, \ldots, x_n$ which are in different sets in both decompositions.

Solution Let us examine an example: $X = \{1, 2, 3, 4, 5, 6, 7, 8, 9\}$ and

$$X = \{1, 2, 3\} \cup \{4, 5, 6\} \cup \{7, 8, 9\} = \{1, 4, 7\} \cup \{2, 3, 6\} \cup \{5, 8, 9\}.$$

We see that $1, 6$, and 9 are in different sets in both decompositions.

In the general case, consider the sets X_i as boys and Y_j as girls. We say that X_i is acquainted with Y_j if $X_i \cap Y_j \neq \emptyset$. Suppose that there is a matching between the boys and the girls such that X_i is matched with $Y_{\sigma(i)}$, for each i. Then we can choose $x_i \in X_i \cap Y_{\sigma(i)}$ and we are done.

In order to prove the existence of such a matching, we will show that Hall's condition holds. Suppose that all the sets have m elements and choose k sets X_i. Their union has mk elements (because the sets are disjoint) and therefore there must be at least k corresponding sets Y_j.

Problem 3.107 A set P consists of 2005 distinct prime numbers. Let A be the set of all possible products of 1002 elements of P, and B be the set of all products of 1003 elements of P. Prove the existence of a one-to-one correspondence f from A to B with the property that a divides $f(a)$ for all $a \in A$.

Solution The set A has $\binom{2005}{1002}$ elements and these are the boys. The set B has the same number of elements, since

$$\binom{2005}{1003} = \binom{2005}{1002}$$

and these are the girls. A boy a is acquainted with a girl b if a divides b. We have to prove that there exists a matching between the boys and the girls. For this, observe that each boy is acquainted with exactly 1003 girls, and each girl is acquainted with exactly 1003 boys. The existence of a matching follows from the observation at the end of the solution of Problem 3.103.

Problem 3.108 The entries of a $n \times n$ table are non-negative real numbers such that the numbers in each row and column add up to 1. Prove that one can pick n numbers from distinct rows and columns which are positive.

Solution Again, the rows are the boys and the columns are the girls. We say that a row is acquainted to a column if the entry at their intersection is a positive number. All we have to do is to show that Hall's condition is fulfilled.

Choose k rows and consider the m columns acquainted to them. Color red the cells of the k rows and blue the cells of the m columns. Consequently, the cells at their intersection will be colored violet. It is not difficult to see that the entries in all red cells are zeroes. Adding up the entries of the k rows yields k, hence the entries at the violet cells add up to k as well. Adding up the entries of the m columns yields m, therefore the sum of entries in the violet and blue cells equals m. Clearly, this implies $k \leq m$, so Hall's condition is indeed fulfilled.

Problem 3.109 There are b boys and g girls present at a party, where b and g are positive integers satisfying $g \geq 2b - 1$. Each boy invites a girl for a dance (of course, two different boys must always invite two different girls). Prove that this can be done in such a way that every boy is either dancing with a girl he knows or all the girls he knows are not dancing.

Solution If Hall's condition is fulfilled, then each boy can invite for the dance a girl he knows. Suppose that Hall's condition is violated and thus, we can find k boys, say $b_1, b_2, \ldots, b_k$, such that the girls they know are $g_1, g_2, \ldots, g_m$, with $m < k$. We choose the maximal k with this property. Now, observe that for the rest of $b - k$ boys and $g - m$ girls, Hall's condition is fulfilled (otherwise the maximality of k is contradicted), hence we can make the $b - k$ boys dance with $b - k$ girls they know. We are left with $g - m - (b - k) \geq 2b - 1 - b + k - m \geq k$ girls and we can make $b_1, b_2, \ldots, b_k$ dance with k of these girls.

Problem 3.110 A $m \times n$ array is filled with the numbers $1, 2, \ldots, n$, each used exactly m times. Show that one can always permute the numbers within columns to arrange that each row contains every number $1, 2, \ldots, n$ exactly once.

Solution Let us show first that we can permute the numbers within columns such that the first row contains every number $1, 2, \ldots, n$ exactly once. Let the columns be the boys and let the numbers $1, 2, \ldots, n$ be the girls. A boy (column) is acquainted with a girl (number) if that number occurs in the column. Now, consider a set of k columns; they contain km numbers, hence there exist at least k distinct numbers among them. Since Hall's condition is fulfilled, there is a matching between the columns and the numbers $1, 2, \ldots, n$. Permuting these numbers to the tops of their respective columns makes the first row contain all n numbers. Finally, a simple inductive argument ends the proof.

Problem 3.111 Some of the AwesomeMath students went on a trip to the beach. There were provided n buses of equal capacities for both the trip to the beach and the ride home, one student in each seat, and there were not enough seats in $n - 1$ buses to fit each student. Every student who left in a bus came back in a bus, but not necessarily the same one.

Prove that there are n students such that any two were on different busses on both rides.

Solution Let the buses be $b_1, b_2, \ldots, b_n$. Denote by X_i the set of students traveling in bus b_i to the beach and by Y_i the set of students traveling in bus b_i on the ride home. Let the X_i's be the boys and the Y_i's be the girls. We say that a boy X_i is acquainted with the girl Y_j if $X_i \cap Y_j \neq \emptyset$. Now consider a set of k boys $X_1, X_2, \ldots, X_k$. If the set of girls acquainted with them has less than k elements then the students in $X_1 \cup X_2 \cup \cdots \cup X_k$ fit in $k - 1$ buses, hence all students fit in $n - 1$ buses, contradicting the hypothesis. Therefore there is a matching between the two sets. If X_i is matched with $Y_{\sigma(i)}$, then we can pick a student from each $X_i \cap Y_{\sigma(i)}$ and we are done.

Glossary

Centroid of a triangle Point of intersection of the medians.

Ceva's theorem and its trigonometric form Let AA', BB', and CC' be three cevians of triangle ABC. Then AA', BB', and CC' are concurrent if and only if either

$$\frac{A'B}{A'C} \cdot \frac{B'C}{B'A} \cdot \frac{C'A}{C'B} = 1$$

or

$$\frac{\sin \angle A'AB}{\sin \angle A'AC} \cdot \frac{\sin \angle B'BC}{\sin \angle B'BA} \cdot \frac{\sin \angle C'CA}{\sin \angle C'CB} = 1.$$

Cevian Any segment joining the vertex of a triangle to a point on the opposite side.

Circumcenter Center of the circumscribed circle.

Circumcircle Circumscribed circle.

Convex quadrilateral The quadrilateral $ABCD$ is convex if the line segments AC and BD intersect (or, equivalently, if all its interior angles are less than $180°$).

Cyclic polygon Polygon that can be inscribed in a circle.

Eisenstein's criterion Let $f = a_0 + a_1 X + \cdots + a_n X^n$ be an integer polynomial. If p is a prime number such that p divides $a_0, a_1, \ldots, a_{n-1}$, p does not divide a_n and p^2 does not divide a_0, then f cannot be written as a product of two (nonconstant) integer polynomials.

Fermat's little theorem If p is a prime number, then $a^p \equiv a \pmod{p}$ for any integer a.

Homothety A homothety of center O and ratio r is a transformation that maps each point P in the plane to a point P' such that $\overline{OP'} = r\overline{OP}$.

Incenter Center of the inscribed circle.

Incircle Inscribed circle.

Inradius The radius of the incircle.

De Moivre's formula For any angle a and any integer n,

$$(\cos a + i \sin a)^n = \cos na + i \sin na.$$

One-to-one function A function f such that if $x \neq y$ then $f(x) \neq f(y)$.

T. Andreescu, B. Enescu, *Mathematical Olympiad Treasures*,
DOI 10.1007/978-0-8176-8253-8, © Springer Science+Business Media, LLC 2011

Orthocenter The point of intersection of the altitudes of a triangle.

Pithot's theorem A circle can be inscribed in the convex quadrilateral $ABCD$ if and only if $AB + CD = BC + DA$.

Root of unity Solution to the equation $z^n - 1 = 0$.

Symmetry center The point O is a symmetry center of a figure F if for any point $M \in F$, there exists $M' \in F$ such that O is the midpoint of the line segment MM'.

Index of Notations

$\mathbf{Z}$ The set of integers

$\mathbf{Q}$ The set of rational numbers

$\mathbf{R}$ The set of real numbers

$\mathbf{C}$ The set of complex numbers

$[a, b]$ The set of real numbers x such that $a \leq x \leq b$

(a, b) The set of real numbers x such that $a < x < b$

AB The segment AB; also the length of the segment AB

$\overline{AB}$ The vector AB

$[F]$ The area of the figure F

$\lfloor x \rfloor$ The integer part of the real number x

T. Andreescu, B. Enescu, *Mathematical Olympiad Treasures*,
DOI 10.1007/978-0-8176-8253-8, © Springer Science+Business Media, LLC 2011

Index to the Problems

ARML Power Question: 2.30.
Balkan Mathematical Olympiad: 1.77, 2.35, 2.42, 3.35.
British Mathematical Olympiad: 2.68.
Bulgarian Mathematical Olympiad: 3.36.
Dutch Mathematical Olympiad: 3.74.
Gazeta Matematica: 1.12, 1.73, 1.102, 2.64.
International Mathematical Olympiad: 1.24, 2.45, 3.24.
Korean Mathematical Olympiad: 3.77.
Kvant: 2.58, 2.59.
Revista Matematica din Timisoara: 1.22, 2.43.
Romanian Mathematical Olympiad: 1.3, 1.25, 1.42, 1.43, 1.45, 1.58, 1.59, 1.62, 1.65, 1.66, 1.79, 1.100, 2.41, 2.50, 2.51, 2.52, 3.12, 3.13, 3.14, 3.15, 3.16, 3.17, 3.18, 3.23, 3.39, 3.40, 3.78, 3.85.
Russian Mathematical Olympiad: 1.38, 1.46, 1.50, 1.52, 1.53, 1.54, 1.75, 1.86, 2.29, 2.31, 2.46, 2.49, 2.57, 3.12, 3.2, 3.8, 3.9, 3.79, 3.80, 3.81, 3.82.
Tournament of the Towns: 3.6, 3.7.
USAMO and US selection tests: 1.27, 1.94.

T. Andreescu, B. Enescu, *Mathematical Olympiad Treasures*,
DOI 10.1007/978-0-8176-8253-8, © Springer Science+Business Media, LLC 2011

CPSIA information can be obtained at www.ICGtesting.com
Printed in the USA
LVOW102124240512

283242LV00004B/8/P